BOUNDARY BEHAVIOR OF HOLOMORPHIC FUNCTIONS OF SEVERAL COMPLEX VARIABLES

BY

E. M. STEIN

PRINCETON UNIVERSITY PRESS

AND

UNIVERSITY OF TOKYO PRESS

PRINCETON, NEW JERSEY

1972

L.C. Card: 71-183062

I.S.B.N.: 0-691-08109-3

Published in Japan exclusively by
University of Tokyo Press;
in other parts of the world by
Princeton University Press

Printed in the United States of America

To Jeremy and Karen

BOUNDARY BEHAVIOR OF HOLOMORPHIC FUNCTIONS OF SEVERAL COMPLEX VARIABLES

By

E. M. Stein

Preface

These are the lecture notes of a course given at Princeton University during the Spring term 1970. The main novel results described here were announced in an earlier note [20]. It is my pleasure to thank the auditors of the course whose participation enriched this project. I want to stress particularly my appreciation to C. L. Fefferman and J. J. Kohn for several valuable suggestions that have been incorporated in the text, to W. Beckner who took notes of the lectures and helped greatly in the preparation of the final text, and last but not least to Miss Elizabeth Epstein for her excellent typing of the manuscript.

Introduction

In classical function theory of one complex variable there is a very close connection between the boundary behavior of holomorphic functions in a domain, and the corresponding problem for harmonic functions in that domain. As a consequence there is one "potential theory" (that of the Laplacian, $\frac{\partial^2}{\partial x^2} + \frac{\partial^2}{\partial y^2}$) which is a fundamental tool for all domains. In the case of more than one complex variable this is no longer so. In the general context the appropriate potential theory (insofar as there is one) should depend on the particular domain considered, and ought, more precisely, to reflect the interplay of the domain with the complex structure of the ambient space $\mathbb{C}^n$.

This point may be understood as follows. Suppose $\mathcal{D}$ is a smooth domain in $\mathbb{C}^n$; it may then also be viewed as a domain in $\mathbb{R}^{2n}$. Now from the second point of view, at the tangent space of a point of $\partial\mathcal{D}$ all directions are essentially equivalent (in the usual potential theory of $\mathbb{R}^{2n}$). However, looked at from $\mathbb{C}^n$, not all directions are equivalent, and there is a natural splitting of this tangent space as a direct sum of a one-dimensional real subspace (the "classical directions"), and a 2n-2 dimensional real subspace (which carries an induced complex structure) of "complex tangential" directions. This splitting explains the distinction between the non-tangential approach of the usual potential theory, and the broader "admissible" approach in the complex case, as we shall see; moreover this splitting is present, in some form, in all notions considered below.

To put matters in still another way: the study of the behavior of holomorphic functions in $\mathcal{D}$ should proceed, in principle, in terms of the basic invariant objects attached to the domain $\mathcal{D}$; the Bergman kernel and its metric, the Szegö kernel, and the Poisson-Szegö kernel, since all these would naturally take into account the simple geometrical considerations just discussed. This is indeed the case of the unit ball in $\mathbb{C}^n$, and there the explicit knowledge of these objects and their interrelation allowed Koranyi [10] to study the complex ball's invariant potential theory. However in the general context not enough is known about these domain functions and so we must use a different approach.

Our results are of two types. In chapter II we obtain an analogue of Fatou's theorem for bounded domains in $\mathbb{C}^n$ with smooth boundary - without making use of any assumptions of pseudo-convexity. The substitute for the Poisson-Szegö kernel is a majorization of pluri-subharmonic functions in $\mathcal{D}$ in terms of a maximal function of their boundary values. This substitute is obtained by first using the standard potential theory of $\mathbb{R}^{2n}$, and then sub-harmonicity in the complex tangential directions. A rather complicated refinement of these arguments leads to an extension of these results to the Nevanlinna class.

In chapter III we introduce the basic notion of a "preferred metric" and carry out the rudiments of potential theory with respect to the Laplace-Beltrani operator of this metric. The existence of such a metric is intimately connected with the strict pseudo-convexity of the domain - and so we operate with this assumption throughout chapter III. It is a natural supposition that the Bergman metric of a strictly pseudo-convex bounded domain with smooth boundary is a preferred metric in our sense. We do not know a proof of this conjecture, but this is not an obstacle since any preferred metric does the job. Our results are here: the local analogue of Fatou's theorem, and the characterization at almost all boundary points of the existence of admissible limits in terms of the finiteness of an analogue of the "area integral". It may be expected that preferred metrics will be useful in other problems as well.

Chapter I is introductory. It contains a sketchy review of some selected topics to give the necessary background and supply some of the motivation for the presentation in the two succeeding chapters.

Table of Contents

Chapter I, first part: Review of potential theory in $\mathbb{R}^N$

This chapter contains a brief review of known facts from potential theory in $\mathbb{R}^N$ and several complex variables.

1. Green's function and Poisson kernel for domains in $\mathbb{R}^N$

Let $\mathcal{D}$ be a bounded smooth domain in $\mathbb{R}^N$. Smooth will mean that the boundary is of class C^2. (See also the discussion in §2 below.) In what follows in this chapter the class $C^{1+\varepsilon}$ would suffice - with slight modifications of the argument. The results would become essentially more difficult if the boundary were only C^1 (or more generally satisfy a Lipschitz condition). However, since many of the applications to complex analysis require a "parabolic" approach, and the definition of pseudo-convexity involves essentially two derivatives, we will restrict consideration to C^2 boundaries.

Let $G(x,y)$ be the Green's function for $\mathcal{D}$, defined in $\mathcal{D} \times \overline{\mathcal{D}}$ -{diagonal}. It is uniquely determined by the following properties: it is smooth on $\mathcal{D} \times \overline{\mathcal{D}}$ -{diagonal}, (i.e. at least of class $C^{2-\varepsilon}$) ; $\Delta_y G(x,y) = 0$ for $x \neq y$; $G(x,y) - c_N|x-y|^{-N+2}$ is harmonic in $y \in \mathcal{D}$, for each fixed x, and $G(x,y)\big|_{y \in \partial\mathcal{D}} = 0$.

Suppose now that $y \in \partial\mathcal{D}$ and ν_y denotes the outward unit normal to $\partial\mathcal{D}$ at y. The function $P(x,y) = -\dfrac{\partial G(x,y)}{\partial \nu_y}$, defined in $\mathcal{D} \times \partial\mathcal{D}$ is the Poisson kernel for $\mathcal{D}$. By the use of Green's theorem and the maximum principle for harmonic functions one can then prove the following known properties of the Poisson kernel, with $x \in \mathcal{D}$ and $y \in \partial\mathcal{D}$:

1) $\qquad P(x,y) \geq c_x > 0 \; .$

2) $\qquad P(x,y) \leq c(\delta(x))^{-N} \; .$

3) $\qquad P(x,y) \leq c \; \dfrac{\delta(x)}{|x-y|^N} \; .$

4) $\qquad$ If $\;$ u $\;$ is harmonic in $\mathcal{D}$ and continuous in $\overline{\mathcal{D}}$ then

$$u(x) = \int_{\partial \mathcal{D}} P(x,y)u(y)d\sigma(y) \qquad .$$

Here $\delta(x)$ denotes the distance of x from $\partial\mathcal{D}$, and $d\sigma(y)$ is the induced Euclidean measure on $\partial\mathcal{D}$. Inequalities 2) and 3) are most conveniently obtained by comparing P with the explicitly known Poisson kernel for the exterior of a ball tangent to $\partial\mathcal{D}$ at y . For further details and references, see e.g. Aronszajn and Smith [1].

2. Boundaries

As we have said before $\mathcal{D}$ will be assumed to be of class c^2. This means that there exists a real-valued function λ defined in a neighborhood of $\overline{\mathcal{D}}$ so that: λ is of class c^2 , $\lambda(x) < 0$ if and only if $x \in \mathcal{D}$, $\{\lambda(x) = 0\} = \partial\mathcal{D}$, and $|\nabla\lambda(x)| > 0$ if $x \in \partial\mathcal{D}$ (The last condition is equivalent with $\left.\dfrac{\partial\lambda}{\partial\nu_x}\right|_{\partial\mathcal{D}} \geq c > 0$, where $\dfrac{\partial}{\partial\nu}$ is the derivative with respect to the outward normal.)

A function λ of the above type will be called a __characterizing function__ for the domain $\mathcal{D}$. Of course there are infinitely many such characterizing functions. Each characterizing function determines a family of approximating subdomains $\mathcal{D}_\varepsilon$ as follows: $\mathcal{D}_\varepsilon = \{x: \lambda(x) < -\varepsilon\}$. Their boundaries $\partial\mathcal{D}_\varepsilon$ are then the level surfaces $\{x: \lambda(x) = -\varepsilon\}$, and for ε sufficiently small and positive $\lambda(x)+\varepsilon$ is a characterizing function for $\mathcal{D}_\varepsilon$.

Once a C^2 domain has been given by its characterizing function, it is technically convenient to allow the wider class of characterizing functions which define it, but which are only assumed to be of class C^1. These are then two particularly noteworthy examples:

(1) $\lambda(x) = -\delta(x)$, for x near $\partial\mathcal{D}$, $x \in \mathcal{D}$.

$= \delta(x)$ for $x \in {}^c\mathcal{D}$, x near $\partial\mathcal{D}$.

δ is the distance of x from $\partial\mathcal{D}$.

Then $\left.\frac{\partial\delta}{\partial\nu}\right|_{\partial\mathcal{D}} = 1$, since distance is measured along the normal.

(2) $\lambda(x) = -G(x^0,x)$, where x^0 is a fixed point in $\mathcal{D}$.

Then $\left.\frac{\partial\lambda}{\partial\nu}\right|_{\partial\mathcal{D}} = P(x^0,x)$.

3. Lemma for harmonic functions

It is useful to know that certain classes of harmonic function on $\mathcal{D}$ do not depend on the characterizing function used to define $\mathcal{D}$. This is contained in the following lemma.

Lemma. Let λ_1 and λ_2 be two characterizing functions of $\mathcal{D}$. Let $\partial\mathcal{D}^i_\varepsilon = \{x: \lambda_i(x) = -\varepsilon\}$, $i = 1,2$. Then for each p, $p \geq 1$ and each harmonic function u in $\mathcal{D}$ the two conditions

(3.1) $$\sup_{\varepsilon>0} \int_{\partial\mathcal{D}^i_\varepsilon} |u(x)|^p d\sigma^i_\varepsilon(x) < \infty, \qquad i = 1,2,$$

are equivalent. ($d\sigma_\varepsilon^i$ is the induced measure on $\partial \mathcal{O}_\varepsilon^i$.)

Proof. It suffices to show that the condition (3.1) for $i=1$ implies the same condition for $i=2$. Now there exist positive constants c, c_1, and c_2 so that if $x \in \partial \mathcal{O}_\varepsilon^2$, and $B(x,c\varepsilon)$ is the ball of radius $c\varepsilon$ centered at x, then

$$B(x,c\varepsilon) \subset \{x:\ -c_1\varepsilon < \lambda_1(x) < -c_2\varepsilon\} .$$

By the mean-value property

$$|u(x)|^p \leq c_3\varepsilon^{-N} \int \chi_\varepsilon(x,y)|u(y)|^p dy ,$$

where χ_ε is the characteristic function of the ball $B(x,c\varepsilon)$. Thus

$$\int_{\partial \mathcal{O}_\varepsilon^2} |u(x)|^p d\sigma_\varepsilon^2(x) \leq c_3\varepsilon^{-N} \int \left| \int_{\partial \mathcal{O}_\varepsilon^2} \chi_\varepsilon(x,y) d\sigma_\varepsilon^2(\lambda) \right| |u(y)|^p dy .$$

However $\int_{\partial \mathcal{O}_\varepsilon^2} \chi_\varepsilon(x,y) d\sigma_\varepsilon^2(x) = 0$ if y is not in the layer $L_\varepsilon = \{y:\ -c_1\varepsilon < \lambda(x) < -c_2\varepsilon\}$, while $\int_{\partial \mathcal{O}_\varepsilon^2} \chi_\varepsilon(x,y) d\sigma_\varepsilon^2(x) \leq c_4\varepsilon^{N-1}$, for $y \in L_\varepsilon$. Thus

$$\int_{\partial \mathcal{O}_\varepsilon^2} |u(x)|^p d\sigma_\varepsilon^2(x) \leq c_5\varepsilon^{-1} \int_{L_\varepsilon} |u(y)|^p dy = c_5\varepsilon^{-1} \int_{c_2\varepsilon}^{c_1\varepsilon} \left(\int_{\partial \mathcal{O}_\eta^1} |u(y)|^p d\sigma_\eta^1(y) \right) d\eta$$

$\leq$ constant. Thus the lemma is proved.

4. Characterization of Poisson integrals

The class of harmonic functions in the lemma above can be characterized in terms of the Poisson kernel for $\mathcal{D}$.

We suppose that u is harmonic in $\mathcal{D}$, and λ is a characterizing function for $\mathcal{D}$. $\mathcal{D}_\varepsilon$ will be the resulting approximating regions given by $\mathcal{D}_\varepsilon = \{x: \lambda(x) < -\varepsilon\}$, $\varepsilon > 0$.

Theorem 1. The following properties are equivalent:

(1) $$\sup_{\varepsilon > 0} \left(\int_{\partial \mathcal{D}_\varepsilon} |u(x)|^p d\sigma_\varepsilon(x) \right)^{1/p} < \infty, \quad 1 \leq p$$

(2) $$u(x) = \int_{\partial \mathcal{D}} P(x,y) f(y) d\sigma(y)$$

where $f \in L^p(\partial \mathcal{D})$, if $p > 1$; when $p = 1$, then $f(y)d\sigma(y)$ has to be replaced by a finite Borel measure on $\partial \mathcal{D}$.

(3) $|u(x)|^p$ has a harmonic majorant if $p < \infty$. When $p = \infty$ we assume that u is bounded in $\mathcal{D}$.

Proof. (1) $\Longrightarrow$ (2).

We assume first that $p > 1$. We shall show that there exists an $f \in L^p(\partial \mathcal{D})$, so that u converges to f in the following (weak) sense. Let $\mathcal{D}'_\varepsilon$ be any approximating region (given by the characterizing function λ', which is not necessarily λ). For each $x \in \partial \mathcal{D}_\varepsilon$, let $\pi_\varepsilon(x)$ be the normal projection of x on $\partial \mathcal{D}$ (which is well-defined if ε is sufficiently small). Define f_ε by $f_\varepsilon(\pi_\varepsilon(x)) = u(x)$, $x \in \partial \mathcal{D}'_\varepsilon$. Then the sequence $\{f_\varepsilon\}$, considered as functions on $L^p(\partial \mathcal{D})$, converges weakly to f as $\varepsilon \longrightarrow 0$.

To see this it suffices to restrict consideration to a subdomain $\mathcal{D}_0$ with the following properties: (a) $\mathcal{D}_0$ is of class C^2. (b) $\mathcal{D}_0 \subset \mathcal{D}$. (c) The boundary of $\mathcal{D}_0$ contains an open subset of the boundary of $\mathcal{D}$. (d) The rest of the boundary of $\mathcal{D}_0$ is contained in (the interior of) $\mathcal{D}$. (e) Suppose ν is an outward unit normal to a point which is on the boundary of both $\partial\mathcal{D}$ and $\partial\mathcal{D}_0$, then $\overline{\mathcal{D}}_0 - \varepsilon\nu \subset \mathcal{D}$ for all positive and sufficiently small ε. The picture is as follows:

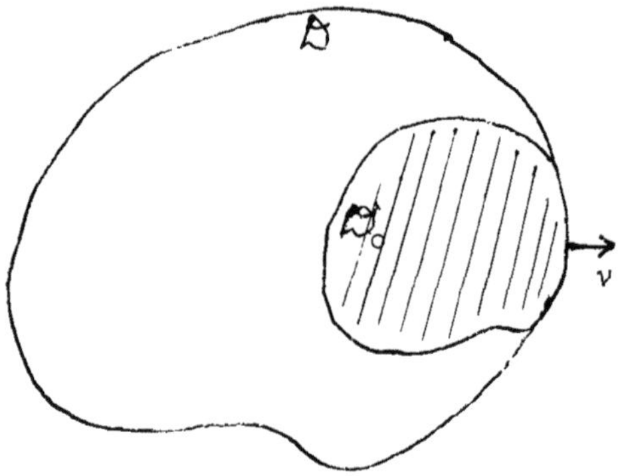

Clearly $\mathcal{D}$ can be covered by finitely many such subdomains. Now u is harmonic in $\mathcal{D}$, and therefore $u(x-\varepsilon\nu)$ is harmonic and continuous in $\overline{\mathcal{D}}_0$. If P_0 is the Poisson kernel for $\mathcal{D}_0$, then

$$u(x-\varepsilon\nu) = \int_{\partial\mathcal{D}_0} P_0(x,y)u(y-\varepsilon\nu)d\sigma^0(y) .$$

Also, in view of the lemma in §3, $\sup\limits_{\varepsilon>0} \int_{\partial\mathcal{D}_0} |u(y-\varepsilon\nu)|^p d\sigma^0(y) < \infty$.

Thus if f is a weak-limit of a subsequence of $\{u(y-\varepsilon\nu)\}_\varepsilon$, then

$$u(x) = \int_{\partial \mathcal{D}_o} P_o(x,y)f(y)d\sigma(y) ,$$

and the desired convergence of u to f follows by standard arguments from the properties of the Poisson kernel listed in section 1. Putting together finitely such sub-domains we get the "weak" convergence of u to f. It remains to be shown that the representation (2) holds with P the Poisson kernel for all of $\mathcal{D}$. For this purpose fix $x^o \in \mathcal{D}$ and set $\lambda(x) = -G(x^o,x)$. If $\varepsilon > 0$, and ε is sufficiently small the domains $\mathcal{D}'_\varepsilon = \{x: \lambda(x) < -\varepsilon\}$ have as their Poisson kernels $P_\varepsilon(x^o,x) = -\frac{\partial G}{\partial \nu_x}(x^o,x)\Big|_{\partial \mathcal{D}_\varepsilon}$. Thus

$$(4.1) \qquad u(x^o) = -\int_{\partial \mathcal{D}_\varepsilon} \frac{\partial G}{\partial \nu_y}(x^o,y)u(y)d\sigma_\varepsilon(y)$$

Now since G is of class $C^{2-\varepsilon}$ in $\mathcal{D} \times \overline{\mathcal{D}}$ -{diagonal}, then

$$-\frac{\partial G}{\partial \nu_y}(x^o, \pi_\varepsilon^{-1}(y)) \longrightarrow -\frac{\partial G}{\partial \nu_y}(x^o,y) = P(x^o,y)$$

uniformly in y; moreover $d\sigma_\varepsilon(y) = \mathcal{J}_\varepsilon(y)d\sigma(\pi_\varepsilon(y))$ where π_ε is the normal projection of $\partial\mathcal{D}'_\varepsilon$ to $\partial \mathcal{D}$, and $\mathcal{J}_\varepsilon$ is the corresponding Jacobian. Also $\mathcal{J}_\varepsilon(y) \longrightarrow 1$ uniformly as $\varepsilon \longrightarrow 0$. Using then the fact that the $u(y)\big|_{\partial \mathcal{D}_\varepsilon}$ tend weakly to f as $\varepsilon \longrightarrow 0$, we get from (4.1) that

$$u(x^o) = \int_{\partial \mathcal{D}} P(x^o,y)f(y)d\sigma(y) ,$$

which is the desired conclusion when $p > 1$. The argument for $p = 1$

is similar except now the weak limits are Borel measures on $\partial\mathcal{D}$ instead of L^p functions.

Proof. (2) $\Longrightarrow$ (3)

This implication is nearly trivial, since if we take $h(x) = \int_{\partial\mathcal{D}} P(x,y)|f(y)|^p d\sigma(y)$, then Hölder's inequality and the fact that $\int_{\partial\mathcal{D}} P(x,y)d\sigma(y) \equiv 1$ shows that h is the desired harmonic majorant of $|u|^p$. A similar argument works if $p = 1$.

Proof. (3) $\Longrightarrow$ (1)

Suppose $|u(x)|^p \leq h(x)$ where h is harmonic in $\mathcal{D}$. We take $\mathcal{D}'_\varepsilon$ as the approximating domains whose boundaries are determined by the level surfaces of the Green's function. Then the formula (4.1) holds for h in place of u. Since $-\frac{\partial G}{\partial \nu_y}(x^o,y)\Big|_{\partial\mathcal{D}'_\varepsilon}$ converges uniformly to $P(x^o,y)$, and $P(x^o,y) \geq c > 0$ (see property 1) in §1) we get that $\sup_{\varepsilon > 0} \int_{\partial\mathcal{D}'_\varepsilon} h(y)d\sigma_\varepsilon(y) < \infty$, which implies $\sup_{\varepsilon > 0} \int_{\partial\mathcal{D}'_\varepsilon} |u(y)|^p d\sigma_\varepsilon(y) < \infty$ and hence on appeal to the lemma in §3 concludes the proof that (3) $\Longrightarrow$ (1).

Small modifications of the proof of Theorem 1 lead to the following corollary:

Corollary. Suppose s is a non-negative, continuous, subharmonic function in $\mathcal{D}$. Then the condition

$$\sup_{\varepsilon > 0} \int_{\partial\mathcal{D}_\varepsilon} (s(x))^p d\sigma_\varepsilon(x) < \infty$$

<u>is equivalent with the existence of a harmonic majorant</u> h <u>of</u> s <u>such that</u> h <u>is the Poisson integral of an</u> L^p <u>function</u> f , <u>when</u> $p > 1$, <u>and</u> h <u>is the Poisson integral of a measure when</u> $p = 1$. <u>Also</u>

$$\|f\|_p \leq \sup_{\varepsilon > 0} \left(\int_{\partial \mathcal{D}_\varepsilon} (s(x))^p d\sigma_\varepsilon(x) \right)^{1/p}, \text{ if } p > 1 .$$

<u>Remark.</u> The proof of the theorem could have been simplified had we used the fact that the Poisson kernels for approximating domains $\mathcal{D}_\varepsilon$ converge (in the appropriate sense) to the Poisson kernel for $\mathcal{D}$. But this fact is not as elementary as the properties 1) to 4) of §1 that we used. See also the literature cited at the end of this chapter.

5. Maximal functions

A key tool in what follows is the use of maximal functions. There were introduced by Hardy and Littlewood and successively generalized and extended by Wiener, Marcinkiewicz and Zygmund and in the context most relevant to by K. T. Smith [18].

It is not our purpose here to give a detailed presentation of these ideas. We shall however formulate a general version of the maximal theorem appropriate for later applications.

Let $\mathcal{M}$ be a measure space (with measure $m(\cdot)$), and suppose that for each point $x \in \mathcal{M}$, there is a family of "balls" $B(x,\rho)$, $0 < \rho < \infty$, with the following properties: there exist positive constants c and K , $c > 1$, so that (α) Each $B(x,\rho)$ is an open bounded set of positive measure.

(β) $B(x,\rho_1) \subset B(x,\rho_2)$, if $\rho_1 \leq \rho_2$.

(γ) $B(x,\rho_1) \cap B(y,\rho_2) \neq \emptyset$ and $\rho_1 \geq \rho_2$ implies that $B(x,c\rho_1) \supset B(y,\rho_2)$.

(δ) $m(B(x,c\rho) \leq Km(B(x,\rho))$.

The simplest example of the above arises if $\mathcal{M}$ is a manifold with a Riemannian metric, and $B(x,\rho)$ is the ball in that metric with center x and radius ρ . However other examples, where the "balls" $B(x,\rho)$ are rather skew, will be decisive later.

For any $f \in L^p(\mathcal{M})$, $\infty \geq p \geq 1$, we can define the maximal function f^* by

$$f^*(x) = \sup_{\rho > 0} \frac{1}{m(B(x,\rho))} \int_{B(x,\rho)} |f(y)|dm(y) .$$

<u>Theorem 2.</u> <u>Suppose</u> $f \in L^p(\mathcal{M})$, $1 \leq p \leq \infty$.

(a) $\|f^*\|_p \leq A_p\|f\|_p$, $1 < p \leq \infty$.

(b) <u>The mapping</u> $f \longrightarrow f^*$ <u>is of weak-type</u> (1.1), <u>i.e.</u>

$m\{x: f^*(x) > \alpha\} \leq \frac{K}{\alpha} \|f\|_1$, <u>if</u> $f \in L^1(\mathcal{M})$.

For a proof of the theorem see the literature cited at the end of this chapter.

The relation between Poisson integrals and maximal functions is both simple and fundamental. We take as before $\mathcal{D}$ to be a bounded domain in $\mathbb{R}^N$ with C^2 boundary $\partial\mathcal{D}$, set $\mathcal{M} = \partial\mathcal{D}$, and we let

$dm = d\sigma$ be the measure induced by Euclidean measure. For $x \in \partial\mathcal{D}$ we write $B(x,\rho) = \{y \in \partial\mathcal{D} : |x-y| < \rho\}$. The properties $\alpha) - \delta)$ are then easily verified for these balls. The resulting maximal function f^* then dominates the "non-tangential" behavior of Poisson integrals.

To be more precise, for each $y \in \partial\mathcal{D}$, and each $\alpha > 0$ we define the "cone" of aperture α and vertex y, $\Gamma_\alpha(y)$ to be $\Gamma_\alpha(y) = \{x \in \mathcal{D} : |x-y| < (1+\alpha)\delta(x)\}$ (where δ is the distance from $\partial\mathcal{D}$). An exact statement is then

<u>Theorem 3.</u> *Suppose* u *is the Poisson integral of* f.

(a) $$|u(x)| \le A_\alpha \sum_{K=1}^{\infty} 2^{-K}(m[B(y,2^K\eta)])^{-1} \int_{B(y,2^K\eta)} |f(z)|d\sigma(z)$$

if $x \in \Gamma_\alpha(y)$ *and* $|x-y| = \eta$.

(b) $$\sup_{x \in \Gamma_\alpha(y)} |u(x)| \le A_\alpha f^*(y).$$

<u>Proof.</u> $$|u(x)| \le \int_{\partial\mathcal{D}} P(x,z)|f(z)|d\sigma(z) = \int_{|z-y|<2\eta} + \sum_{K=2}^{\infty} \int_{2^{K-1}\eta<|z-y|<2^K\eta}.$$

Since $P(x,z) \le c\,\dfrac{\delta(x)}{|x-z|^N}$ and $|x-z| \ge \delta(x)$, we get that $P(x,z) \le c(\delta(x))^{-N+1}$. The cone condition, $|x-y| < (1+\alpha)\delta(x)$, shows that $\delta(x) \ge \dfrac{|x-y|}{1+\alpha} = \dfrac{\eta}{1+\alpha}$. Thus

$$\int_{|z-y|<2\eta} P(x,z)|f(z)|d\sigma(z) \leq c_\alpha \eta^{-N+1} \int_{B(y,2\eta)}$$

$$\leq A_\alpha \cdot \frac{1}{m(B(y,2\eta))} \int_{B(y,2\eta)} |f(z)|d\sigma(z) ,$$

since $m(B(y,2\eta)) \sim c\eta^{N-1}$.

Similarly $|x-y| \geq \delta(x)$, thus $\delta(z) \leq \eta$, and $|x-z| \geq |z-y| - |y-x| \geq 2^{K-1}\eta - \eta \geq 2^{K-2}\eta$, if $K \geq 2$. Hence $P(x,z) \leq c'2^{-NK}\eta^{-N+1}$, whenever $2^{K-1}\eta < |z-y| < 2^K\eta$, and $|x-y| = \eta$ This shows that

$$\int_{2^{K-1}\eta < |z-y| < 2^K\eta} \leq A\, 2^{-K}[m(B(y,2^K\eta))]^{-1} \int_{B(y,2^K\eta)} |f(z)|d\sigma(z) .$$

Upon summing in K we get conclusion (a). Conclusion (b) is then an immediate consequence.

From these estimates it follows by standard arguments that non-tangential limits exist almost everywhere for Poisson integrals. A precise statement is as follows:

Theorem 4. Suppose u is harmonic in $\mathcal{D}$ and is bounded, or more generally (as theorem 1 shows) assume that u is the Poisson integral of and L^p function f , $1 \leq p$. Then the non-tangential limits $\lim_{\substack{x \to y \\ x \in \Gamma_\alpha(y)}} u(x)$, exist for almost every $y \in \partial\mathcal{D}$.

The non-tangential limit exists at every point y^0 which is in the "Lebesgue set" of f , that is for which

$$\frac{1}{mB(y^o,\eta)} \int_{B(y^o,\eta)} |f(y) - f(y^o)|\,d\sigma(y) \longrightarrow 0 \,, \text{ as } \eta \to 0 \,.$$

That almost all y^o satisfy the latter property is of course a consequence of theorem 2. The unproved assertions made here regarding theorem 4 and the properties of the Lebesgue set can be obtained by following very closely the well-known arguments in the case of the unit disc or of the half-space in $\mathbb{R}^N$ (for which see e.g. Stein [19]).

Theorem 4 (in combination with theorem 1) is "Fatou's theorem" for the present context. For further details see K. T. Smith [18] and the other references given at the end of this chapter.

6. Local Fatou theorem and area integral

Theorem 4 has a local version. We require a definition. Suppose $y \in \partial\mathcal{D}$. Then we say that u is <u>non-tangentially bounded at</u> y if u is bounded in $\Gamma_\alpha(y)$, for some cone $\Gamma_\alpha(y)$ with vertex y.

<u>Theorem 5.</u> <u>Suppose</u> u <u>is harmonic in</u> $\mathcal{D}$. <u>Then for almost every point</u> $y \in \partial\mathcal{D}$ <u>the following two statements are equivalent:</u>

(a) u <u>is non-tangentially bounded at</u> y .

(b) u <u>has a non-tangential limit at</u> y .

The question whether u has non-tangential limits can also be answered in terms of the "area integral". For any harmonic function u and a boundary point $y \in \partial\mathcal{D}$, we consider $Su(y)$ defined by

$$Su(y) = \left(\int_{\Gamma_\alpha(y)} |\nabla u|^2 (\delta(x))^{2-N} dx \right)^{1/2} .$$

Here $|\nabla u|^2 = \sum_{j=1}^{N} \left|\frac{\partial u}{\partial x_j}\right|^2$, $\delta(x)$ is the distance of x from $\partial \mathcal{D}$, dx is Euclidean measure on $\mathbb{R}^N$.

Theorem 6. Suppose u is harmonic in $\mathcal{D}$. Then for almost every every point $y \in \partial \mathcal{D}$ the following two statements are equivalent.

(a) u has a non-tangential limit at y .

(b) $Su(y) < \infty$.

These theorems were originally obtained by Privalov, Plessner, Marcinkiewicz and Zygmund, and Spencer in the classical case $N = 2$ by the use of complex-variable techniques. (See e.g. the exposition in Zygmund [26], chapter 14.) Methods which are effective for $N > 2$ go back to Calderón and the author. These matters are carried out the case when $\mathcal{D}$ is a half-space in $\mathbb{R}^N$. The present case, for bounded $\mathcal{D}$ with smooth boundary, can be done in the same way if one makes use of the facts about harmonic functions which are discussed above. The reader may also consult K. O. Widman [25], where detailed proofs of theorems 5 and 6 may be found, together with various generalizations.

Chapter I, second part: Review of some topics in several complex variables

7. Bergman kernel, Szegö kernel, and Poisson-Szegö kernel

We now consider the standard complex n-dimensional Euclidean space $\mathbb{C}^n$. If we were to disregard the complex structure of $\mathbb{C}^n$, keeping only the real structure, we would be led to the usual identification of $\mathbb{C}^n$ with $\mathbb{R}^N$, where $N = 2n$. In terms of this identification holomorphic functions in $\mathbb{C}^n$ are harmonic in $\mathbb{R}^N$, and this explains the relevance of the previous material. Nevertheless for the study of holomorphic functions we need those objects which are intrinsically related to the domain in question and which reflect more intimately the complex structure of $\mathbb{C}^n$.

We begin with the Bergman kernel. For the purposes of this kernel we need assume only that $\mathcal{D}$ is a bounded domain in $\mathbb{C}^n$ (without restriction of smoothness of the boundary). One then defines the well-known Hilbert space $\mathcal{L}^2(\mathcal{D})$ of holomorphic functions f in $\mathcal{D}$ with norm $\|f\| = \left(\int_{\mathcal{D}} |f(z)|^2 d\omega(z)\right)^{1/2} < \infty$. (Here $d\omega(z)$ denotes the Euclidean measure in $\mathbb{C}^n$.) The fact that this space is complete follows from the easily proved inequality

$$\sup_{z \in K} |f(z)| \le A_K \|f\| , \qquad K \text{ a compact subset of } \mathcal{D}. \tag{7.1}$$

Next let $\{\varphi_j(z)\}_{j=1}^{\infty}$ be any orthonormal basis for $\mathcal{L}^2(\mathcal{D})$. It can be shown that the series $\sum_{j=1}^{\infty} \varphi_j(z)\overline{\varphi}_j(\zeta)$ converges uniformly in every compact subset of $\mathcal{D} \times \mathcal{D}$; its sum $K(z,\zeta)$ is in fact independent of the particular choice of the basis, and it is characterized by the following three properties:

(1) $K(z,\zeta) = \overline{K(\zeta,z)}$.

(2) For each fixed $\zeta \in \mathcal{D}$, $K(z,\zeta) \in \mathcal{L}^2(\mathcal{D})$.

(3) $f(z) = \int_{\mathcal{D}} K(z,\zeta)f(\zeta)d\omega(\zeta)$, if $f \in \mathcal{L}^2(\mathcal{D})$.

In view of these facts it is easy to see what happens to the Bergman kernel when we transform one bounded domain biholomorphically into another. Let $\mathcal{D}$ and $\mathcal{D}'$ be two such domains, $K_{\mathcal{D}}$ and $K_{\mathcal{D}'}$ their Bergman kernel functions, and suppose $z' = \psi(z)$ is a biholomorphic mapping of $\mathcal{D}$ onto $\mathcal{D}'$. Then

$$(7.2) \qquad \det \frac{\partial\psi(z)}{\partial z} \; \overline{\det \frac{\partial\psi(\zeta)}{\partial\zeta}} \; K_{\mathcal{D}'}(z',\zeta') = K_{\mathcal{D}}(z,\zeta) .$$

If $\mathcal{D}$ were a homogeneous domain (i.e. with a transitive group of holomorphic self-mappings), then the transformation law (7.2) could be used to determine $K(z,z)$ and thus $K(z,\zeta)$, (in principle) at least up to a multiplicative constant. We shall see example of this later, but we first record another basic fact about the Bergman kernel.

With the use of the Berman kernel we can write down a Hermitian metric form on $\mathcal{D}$ as follows:

$$(7.3) \qquad \begin{cases} ds^2 = \sum_{i,j=1}^{n} g_{ij} dz_i d\overline{z_j} \\ \text{with } g_{ij} = \dfrac{\partial}{\partial z_i \partial \overline{z_j}} [\log K(z,z)] . \end{cases}$$

Theorem 7. (1) The Bergman metric (7.3) is positive definite.

(2) Suppose $\mathcal{D}$ and $\mathcal{D}'$ are two bounded domains. Then every biholomorphic mapping of $\mathcal{D}$ to $\mathcal{D}'$ is an isometry in terms of the respective Bergman metrics of $\mathcal{D}$ and $\mathcal{D}'$.

Part (1) of the theorem is essentially a consequence of the fact that the logarithm of the absolute value of an analytic function is pluri sub-harmonic. Part (2) follows easily from the transformation (7.2).

Example. An easily computable example arises when $\mathcal{D}$ = unit disc in $\mathbb{C}^1$. Then $K(z,\zeta) = \frac{1}{\pi}\frac{1}{(1-z\cdot\bar{\zeta})^2}$ and the Bergman metric is

$$ds^2 = \frac{2dzd\bar{z}}{(1-|z|)^2} = \frac{2dx^2+dy^2}{(1-|z|^2)^2},$$ which is of course the Poincaré metric.

We turn next to the Szegö kernel, whose definition is similar to a certain extent to that of the Bergman kernel, except that the integration is now taken on the boundary $\partial\mathcal{D}$, instead of over $\mathcal{D}$. Here it will be again necessary to assume that $\partial\mathcal{D}$ satisfies the smoothness condition (class C^2) imposed earlier.

We consider the harmonic functions in $\mathcal{D}$ (harmonic in the sense of $\mathbb{R}^N$, $N=2n$) which are Poisson integrals of $L^2(\partial\mathcal{D})$ functions, and which are holomorphic in $\mathcal{D}$. That the space of boundary values is a closed subspace of $L^2(\partial\mathcal{D})$ follows immediately from the inequality

(7.1)' $$\sup_{z\in K}|u(z)| \leq A_K\|f\|_2, \quad K \text{ a compact subset of } \mathcal{D},$$

which itself is a direct consequence of theorem 1 and 2 in §1. We may restate matters as follows: We define $H^2(\mathcal{D})$ to be the space of function $f(z)$, holomorphic in $\mathcal{D}$, for which

$$\sup_{\varepsilon > 0} \int_{\partial\mathcal{D}_\varepsilon} |f(z)|^2 d\sigma_\varepsilon(z) < \infty .$$

In view of theorem 1 each such f is the Poisson integral of a boundary function (which we denote by $f(\zeta)$). We norm the space with the norm $\|\cdot\|_2$ as follows, $\|f\|_2 = \left(\int_{\partial\mathcal{D}} |f(\zeta)|^2 d\sigma(\zeta)\right)^{1/2}$. With this norm (and by the use of inequality (7.1)') we see that $H^2(\mathcal{D})$ is a complete Hilbert space.

Now let $\{\psi_j(z)\}$ be an orthonormal basis for $H^2(\mathcal{D})$. Then, as in the case of the Bergman kernel, the series

$$S(z,\zeta) = \sum_{j=1}^{\infty} \psi_j(z)\overline{\psi_j(\zeta)}$$

converges uniformly for z,ζ restricted to any compact subset of $\mathcal{D} \times \mathcal{D}$; the sum is independent of the particular choice of the basis; for each fixed $z \in \mathcal{D}$, $\overline{S(z,\zeta)} \in H^2(\mathcal{D})$, as a function of ζ, and S satisfies the reproducing property:

$$f(z) = \int_{\partial\mathcal{D}} S(z,\zeta)f(\zeta)d\sigma(\zeta) , \qquad z \in \mathcal{D}, \qquad f \in H^2(\mathcal{D}) .$$

We are now in a position to define a "Poisson kernel" intimately associated with this Szegö kernel. It is necessary to emphasize that there are important differences between this Poisson kernel (which we shall write as $\mathcal{P}$) and the Poisson kernel P, arising from the potential theory in $\mathbb{R}^N$.

We set $\mathcal{P}(z,\zeta) = \frac{|S(z,\zeta)|^2}{S(z,z)}$, for $z \in \mathcal{D}$, $\zeta \in \partial\mathcal{D}$. The basic properties of $\mathcal{P}$ are as follows:

Theorem 8. (1) $\mathcal{P}(z,\zeta) \geq 0$

(2) $f(z) = \int_{\partial\mathcal{D}} \mathcal{P}(z,\zeta)f(\zeta)d\sigma(\zeta)$,

whenever f is holomorphic in $\mathcal{D}$ and continuous in $\overline{\mathcal{D}}$.

To prove (2) it is only necessary to invoke the reproducing property for the Szezö kernel at the fixed point z with the analytic function F, where $F(\zeta) = \overline{S(z,\zeta)}f(\zeta)$. Notice that $F \in H^2(\mathcal{D})$.

The definition of the Poisson kernel $\mathcal{P}$ and theorem 8 raise several questions.

Question 1. Does $\mathcal{P}$ give an approximation to the identity, in the sense that if $f(\zeta)$ is continuous on $\partial\mathcal{D}$, then $u(z) = \int_{\partial\mathcal{D}} \mathcal{P}(z,\zeta)f(\zeta)d\sigma(\zeta)$ is continuous in $\overline{\mathcal{D}}$ and $u|_{\partial\mathcal{D}} = f$? The answer to this question is in the affirmative if every point $\zeta \in \partial\mathcal{D}$ satisfies the strict maximum property: there exists a function f, holomorphic in $\mathcal{D}$ and continuous in $\overline{\mathcal{D}}$, for which $|f(\zeta)|$ is a strict maximum. For the connection of this maximum property with pseudo-convexity of $\mathcal{D}$, see Gunning and Rossi [4], p. 275.

Question 2. Does the reproducing property of $\mathcal{P}$ (property (2) of theorem 8) hold for every element of $H^2(\mathcal{D})$? Let $H^2_0(\mathcal{D})$ be the closure in $H^2(\mathcal{D})$ of the holomorphic functions which are continuous in $\overline{\mathcal{D}}$. It suffices then to know that $H^2_0(\mathcal{D}) = H^2(\mathcal{D})$. Whether this holds in general seems not easy to decide. However when $\mathcal{D}$ is strictly pseudo-convex one can prove, using estimates for the $\bar{\partial}$ problem, that $H_0^2 = H^2$ and thus in that case the answer is affirmative also. (The relevant estimates are implicitly contained in Kohn [9].)

Question 3. What are the relations between K and S, and between $\mathcal{P}$ and the Poisson kernel P of §1? When $n = 1$ much is known, and K, S, $\mathcal{P}$ and P are closely connected. For example when $\mathcal{D}$ is the unit disc

$$K(z,\zeta) = \frac{1}{\pi}\frac{1}{(1-z\bar{\zeta})^2}, \quad S(z,\zeta) = \frac{1}{2\pi}\frac{1}{(1-z\bar{\zeta})}, \text{ and}$$

$$\mathcal{P}(z,\zeta) = P(z,\zeta) = \frac{1}{2\pi}\cdot\frac{1-|z|^2}{|1-z\bar{\zeta}|^2}.$$

Significant relations persist for more general domains when $n = 1$. In particular if $\mathcal{D}$ is simply connected, then $\mathcal{P} = P$. For these facts see Bergman [2].

However when $n \geq 2$ the situation changes substantially. Thus $\mathcal{P}$ can never be expected to equal P, since their singularities are of a different nature, as we shall see below. The relation of K and S is known also only in very special circumstances. The case where our knowledge of all these kernels is explicit is that of the unit ball in $\mathbb{C}^n$. For this reason, and because it affords valuable hints for the more general case, we turn to some detailed computations for the unit ball.

8. The unit ball in $\mathbb{C}^n$

Let $\mathbb{B}^n$ denote the unit ball in $\mathbb{C}^n$, $\mathbb{B}^n = \{z = (z_1,\dots,z_n): |z| < 1\}$. $\mathbb{B}^n$ has a transitive group of holomorphic self-mappings - the generalization of the fractional linear transformations when $n = 1$. To describe these we consider an auxilliary vector space $\mathbb{C}^{n+1}$, with points $t = (t_1,t_2,\dots,t_{n+1})$ and the indefinite hermitian form $\mathcal{F}(t,t) = -\sum_{j=1}^{n} |t_j|^2 + |t_{n+1}|^2$.

We consider the complex-linear transformations which preserve $\mathcal{F}$. It is convenient to write these $(n+1)\times(n+1)$-matrices in block form,

$$g = \begin{pmatrix} A & B \\ C & D \end{pmatrix}$$

where $A = n\times n$ matrix, $B = n\times 1$ matrix, $C = 1\times n$ matrix, $D = 1\times 1$ matrix. If $\mathcal{J}$ is the matrix with $A = -I$, $B = 0$, $C = 0$, $D = 1$, then g preserves $\mathcal{F}$ if and only if $g'\mathcal{J}\bar{g} = \mathcal{J}$, where $'$ and $^-$ denote respectively the transpose and complex conjugate. The conditions on g are equivalent to the following pair of sets of identities:

$$\begin{cases} A'\bar{A} - C'\bar{C} = I \\ |D|^2 - B'\bar{B} = 1 \\ A'\bar{B} = C'\bar{D} \end{cases} \quad \text{or} \quad \begin{cases} \bar{A}A' - \bar{B}B' = I \\ |D|^2 - \bar{C}C' = 1 \\ \bar{A}C' = \bar{B}D \end{cases}$$

Now the set of t's where $\mathcal{F}(t,t) > 0$, is equivalent via the transformation $z_1 = t_1/t_{n+1}$, $z_2 = t_2/t_{n+1},\dots,z_n = t_n/t_{n+1}$, to the set of z's, where $|z| < 1$, i.e. the unit ball in $\mathbb{C}^n$. Thus each such g induces a biholomorphic mapping

$$(8.1) \qquad w: z \longrightarrow (Az+B)/(Cz+D)$$

to itself (where z is regarded as an $n \times n$ matrix). To see that the resulting group is transitive it suffices to show that one can map each fixed z^o, $|z^o| < 1$ to 0. Now it is easy to see that $1-(z^o)'\bar{z}^o$ is a positive number and $I - \bar{z}^o(z^o)'$ is a positive definite $n \times n$ matrix. Let R and Q be respectively 1×1 and $n \times n$ matrices which satisfy $\bar{R}(1-(z^o)'\bar{z}^o)R = 1$, $\bar{Q}(1-\bar{z}^o(z^o)')Q' = 1$. Then

$$(8.2) \qquad z \longrightarrow Q(z-z^o)/[1-(\bar{z}^o)'z]R = \varphi(z)$$

maps z^o to z and is of the form (8.1) with $A = Q$, $B = -Qz^o$, $C = -(z^o)'R$, $D = R$. The subgroup of these mappings leaving the origin fixed arises when A is a unitary matrix, D has absolute value 1 and B and C vanish, and is identical with $U(n) \times U(1)$. Since the full group defined above is the "unitary" group of the hermitian form $\mathcal{F}$, (i.e. $U(n,1)$) we have therefore a natural identification of the unit ball with

$$U(n,1)/U(n) \times U(1) .$$

We can now compute the Bergman kernel for the unit ball. Let $K(z) = K(z,z)$, and $\mathcal{J}_\varphi$ denote Jacobian of the mapping (8.2) from z^o to 0. Because of the transformation law (7.2) where $K(z^o) = K(0)|\mathcal{J}_\varphi|^2$. However a simple computation shows

$$\mathcal{J}_\varphi = (1-|z^o|^2)^{-1}R^{-1}Q \text{ , thus}$$

$$|\mathcal{J}_\varphi|^2 = \frac{1}{(1-|z^o|^2)^{n+1}} \text{ , and hence}$$

$$K(z^o) = \frac{K(0)}{(1 - |z^o|^2)^{n+1}} \,.$$

It is easy to verify that $K(0)$ is the reciprocal of the volume of the unit ball and hence

(8.3) $$K(z,\zeta) = \frac{n!}{\pi^n} \cdot \frac{1}{(1 - z\cdot\overline{\zeta})^{n+1}} \qquad (*)$$

where $z\cdot\overline{\zeta} = z_1\cdot\overline{\zeta}_1 + z_2\cdot\overline{\zeta}_2 \cdots + z_n\cdot\overline{\zeta}_n$.

To obtain the Bergman metric $ds^2 = \Sigma\, g_{ij} dz_i d\overline{z}_j$ we calculate $g_{ij} = - \frac{\partial}{\partial z_i \partial \overline{z}_j} \log (1 - |z|^2)^{n+1}$. The result is

(8.4) $$g_{ij} = \frac{n+1}{(1 - |z|^2)} [(1 - |z|^2)\delta_{ij} + \overline{z}_i z_j] \,.$$

Written in matrix notation the metric is given by

$$ds^2 = \frac{n+1}{1 - |z|^2} (dz)'[1 - \overline{z}z']^{-1}(\overline{dz})$$

while the matrix g^{ij} inverse to g_{ij} is given by

(8.5) $$(g^{ij}) = \left(\frac{1 - |z|^2}{n+1}\right) [1 - \overline{z}z'] \,.$$

The Szegö kernel for the unit ball can be calculated in a somewhat similar way.

(*) Footnote: We use here the fact that if $K(z,\zeta)$ is holomorphic in $z \in \mathcal{D}$ and antiholomorphic in $\zeta \in \mathcal{D}$, then $K(z,z)$ uniquely determines K .

By an explicit calculation, if $w(z) = Az+B/Cz+D$ then

$$1 - |w|^2 = \frac{1 - |z|^2}{|Cz+D|^2} . \tag{8.6}$$

Also the absolute value of the Jacobian determinant of $w(z)$, $|\frac{\partial w(z)}{\partial z}|$, equals

$$|\det(B\bar{z}' + A)|^{-(n+1)} = |Cz + D|^{-(n+1)} .$$

Because of (8.6), $d\sigma(w) = |Cz + D|^2 \, |\frac{\partial w}{\partial z}|^2 d\sigma(z)$, that is,

$$d\sigma(w) = |Cz + D|^{-2n} d\sigma(z) ,$$

where $d\sigma$ denotes the measure on the boundary of the unit ball. Thus $S(z,\zeta)$ satisfies the transformation law

$$S(z,\zeta) = S(w(z),w(\zeta))(Cz + D)^{-n}(\overline{C\zeta + D})^{-n} .$$

In particular, if we take $z = \zeta = z^0$, and w to be the transformation (8.2) we get, since $\mathcal{J}_\varphi = \frac{\partial w}{\partial z}$,

$$S(z^0,z^0) = S(0,0)(1 - |z^0|^2)^{-n}$$

and $S(0,0)$ is the reciprocal of the "area" of the sphere. Hence

$$S(z,\zeta) = \frac{(n-1)!}{2\pi^n} \, \frac{1}{(1 - z\cdot\bar{\zeta})^n} . \tag{8.7}$$

The resulting Poisson-Szegö kernel is then

$$\mathcal{P}(z,\zeta) = \frac{(n-1)!}{2\pi^n} \, \frac{(1 - |z|^2)^n}{|1 - z\cdot\bar{\zeta}|^{2n}} . \tag{8.8}$$

This should be compared to the Poisson kernel $P(z,\zeta)$ discussed in §§1 - 6 for the classical potential theory (here in the case of the unit ball of $\mathbb{R}^N$, with $N = 2n$). Then

$$P(z,\zeta) = \frac{(n-1)!}{2\pi^n} \cdot \frac{1-|z|^2}{|z-\zeta|^{2n}} \ .$$

The computation sketched in this section can be found in complete detail - together with various generalizations - in the monographs of Siegel [17] and Hua [8]. The reader should also consult those works and the literature cited at the end of this chapter for the proofs of the following assertions which clarify the roles of the Bergman, Szegö, and Poisson kernels for the unit ball.

Assertion 1.

Let us be given a Riemannian manifold with metric

$$ds^2 = \sum_{i,j} h_{ij} dt_i dt_j \ .$$

Let h^{ij} be the matrix inverse to h_{ij}. For any smooth function grad f is the vector field which in local coordinates is given by $\text{grad } f = \Sigma\, h^{ij} \frac{\partial f}{\partial t_i} \frac{\partial}{\partial t_j}$. For each vector field X, div X is the scalar function given in local coordinates by

$$\text{div } X = \frac{1}{\sqrt{h}} \Sigma \frac{\partial}{\partial t_j} (\sqrt{h}\, X_j)$$

where h denotes the determinant of (h_{ij}). The Laplace-Beltrami operator, Δ, is then defined by

$$\Delta f = \text{div (grad } f) \ .$$

The fact that is asserted is: Δ commutes with isometries of the underlying manifold.

For the case of the unit ball in $\mathbb{C}^n$ with its Bergman metric

$$ds^2 = \Sigma\, g_{ij} dz_i d\bar{z}_j$$

the fractional linear transformations coming from $U(n,1)$ are isometries, and so Δ commutes with the action of this group.

Incidentally an explicit form for Δ in the case of a hermitian metric is

$$\Delta f = \frac{2}{g} \sum_{i,j} \left\{ \frac{\partial}{\partial \bar{z}_i} \left(g g^{ij} \frac{\partial f}{\partial z_j} \right) + \frac{\partial}{\partial z_j} \left(g g^{ij} \frac{\partial f}{\partial \bar{z}_i} \right) \right\}$$

where g^{ij} is the inverse matrix to g_{ij} and g is the determinant of the latter matrix.

<u>Assertion 2.</u>

We now suppose that the hermitian metric is also Kähler. This means that the two-form $\Sigma\, g_{ij} dz_i \wedge d\bar{z}_j$ is closed. "Locally" this means that there exists a real function, $G(z)$, so that $g_{ij} = \dfrac{\partial^2 G}{\partial z_i \partial \bar{z}_j}$; or equivalently if given any point z^0 we can find a coordinate system (by holomorphic change of variables), so that $g_{ij}(z) = \delta_{ij} + O(|z-z^0|^2)$ for z near z^0. (For the latter equivalence see Kohn [9, pp. 129-130].) Our assertion now is: <u>if the metric is Kähler, the Laplace-Beltrami operator does not involve any first order terms and, more particularly, every holomorphic function is annihilated by it.</u> More precisely, we have the formula

$$\Delta f = 4 \sum_{i,j} g^{ij} \frac{\partial^2 f}{\partial \bar{z}_i \partial z_j} \quad . \tag{8.9}$$

Notice that the Bergman metric is always Kähler. In view of the formula (8.5) we get for the case of the unit ball

$$(8.10) \qquad \Delta f = \frac{4}{n+1}(1-|z|^2)\sum_{i,j}(\delta_{ij} - z_i\bar{z}_j)\frac{\partial^2 f}{\partial z_i \partial \bar{z}_j} .$$

The proof of the assertion follows from the fact that for functions f, if the metric is Kähler then $\Delta f = 2\vartheta\bar{\partial}f$, with $\partial f = \Sigma \frac{\partial f}{\partial z_k} dz_k$, $\bar{\partial}f = \Sigma \frac{\partial f}{\partial \bar{z}_k} d\bar{z}_k$, and $\vartheta = -*\partial*$. See e.g. Schiffer and Spencer [16].

<u>Assertion 3.</u> (For the unit ball)

<u>For fixed $\zeta \in \partial \mathcal{B}$, the Poisson-Szegö kernel $\mathcal{P}(z,\zeta)$ is "harmonic" in the sense that it is annihilated by the invariant Laplacian</u> (8.10).

This assertion is the consequence of certain other facts of independent interest. Since $\mathcal{P}(z,w) = \frac{|S(z,w)|^2}{S(z,z)}$, the transformation law for the Szegö kernel leads to the transformation law $\mathcal{P}(w(z),w(\zeta)) = \mathcal{P}(z,\zeta)|Cz+D|^{2n}$, under (8.1). Together with the fact, already noted, that $d\sigma(w) = |Cz+D|^{-2n}d\sigma(z)$, we obtain the following results about the Poisson kernel.

For each fixed z, $|z| < 1$, consider the measure $d\mu_z(\zeta)$ on the boundary, $\{|\zeta| = 1\}$, given by $\mathcal{P}(z,\zeta)d\sigma(\zeta) = d\mu_z(\zeta)$. We have then

$$(8.11) \qquad d\mu_{w(z)}(w(\zeta)) = d\mu_z(\zeta)$$

and thus the passage from a function $f \in L^p(|\zeta| = 1)$, to its Poisson-Szegö integral, $u(z) = \int \mathcal{P}(z,\zeta)f(\zeta)d\sigma(\zeta) = \int f(\zeta)d\mu_z(\zeta)$, commutes with the transformations given by (8.1).

Therefore any observation we can make about a Poisson integral at the origin can be reinterpreted as an analogous statement holding at an arbitrary point z, $|z| < 1$, in view of the transitivity of the group of transformations (8.1). Now $u(0) = \int d\mu_0(\zeta)f(\zeta) = \frac{(n-1)!}{2\pi^n} \int_{|\zeta|=1} f(\zeta)d\sigma(\zeta)$, is the mean value of f on the unit sphere. Hence the value of the Poisson integral at any fixed point z is the integral of f against that measure which is obtained by transforming the normalized invariant measure on the sphere under a fractional linear transformation (8.1) that maps 0 to z.

To go further we notice also that the value of $u(0)$ equals the mean-value of u taken over any sphere whose center is the origin, namely if $0 < \rho < 1$,

(8.12) $$u(0) = \frac{(n-1)!}{2\pi^n} \int_{|\zeta|=1} u(\rho\zeta)d\sigma(\zeta) = \int_K u[k(\rho\zeta^0)]dk$$

where $|\zeta^0| = 1$, and K is the group $U(x) \times U(1)$ of transformations (8.1) keeping the origin fixed, and dk the normalized Haar measure for K. This identity can be written, in view of (8.11), as

$$d\mu_0(\zeta) = \int_K d\mu_{z^0}(k^{-1}\zeta)dk, \qquad \text{with } z^0 = \rho\zeta^0,$$

and can be verified since both sides are the normalized invariant measures (invariant by K) on the surface of the sphere. The final step is to transform (8.12) via an arbitrary element (8.1). This gives the general "mean-value-property" for Poisson-Szegö integrals,

$$(8.13) \qquad u(z) = \int_{K_z} u(\bar{k}(\rho\zeta^o))d\bar{k} .$$

Here K_z is the subgroup of transformations (8.1) leaving z fixed, $|z| < 1$, and $d\bar{k}$ is its normalized Haar measure. Of course $K_z = w^{-1}Kw$, where w is any transformation taking z to the origin. Now it is known, under very general circumstances (see e.g. Helgason [5], chapter 10), that any function satisfying such a mean-value property is annihilated by all invariant differential operators without constant term. Since the Laplace-Beltrami operator (8.10) is such an invariant operator, any Poisson integral is annihilated by it. The above argument also holds for $\mathcal{P}(z,\zeta)$ (with ζ fixed), and so the assertion 3 is proved.

Assertion 4. (For the unit ball)

We want to describe the analogue of the non-tangential domination in theorem 3(b), section 5, in the context of Poisson-Szegö integrals for the unit ball.

In view of the explicit formula (8.8) for the Poisson-Szegö kernel we may expect that the appropriate family of "balls" on the surface of the sphere are the ones defined by

$$B(\zeta^o,\rho) = \{|\zeta| = 1: 1-\zeta^o\cdot\bar{\zeta}^o < \rho\} , \qquad |\zeta^o| = 1 .$$

We shall show below that if $m(\cdot)$ is the usual measure on the surface of the sphere, then these balls satisfy the properties α) to δ) in §5.

Thus for any $f \in L^p$ we may define f^* in terms of these balls. Similarly for any ζ^o on the boundary $(|\zeta^o| = 1$, we define the approach region

$$\mathcal{A}_\alpha(\zeta^o) = \{|z| < 1, |1 - z\bar{\zeta}^o| < \alpha\delta(z)\}$$

where $\delta(z) = 1 - |z|$ is the distance of z from the unit sphere. Let now

$$u(z) = \int_{|\zeta| = 1} \mathcal{P}(z,\zeta) f(\zeta) d\sigma(\zeta) .$$

Our assertion is:

(8.14) $$\sup_{z \in \mathcal{A}_\alpha(\zeta)} |u(z)| \leq A_\alpha f^*(\zeta) , \qquad |\zeta| = 1 .$$

For a proof of this result, and also the analogue of theorem 3, part (a), see Koranyi [10].

Let us examine (8.14). For simplicity, assume that $\zeta^o = (1,0,\ldots,0)$. Then the balls $B(\zeta^o,\rho)$ are essentially given by $\{\zeta: |\text{Im}\ \zeta_1| < \rho, \sum_{K=2}^{n} |\zeta_K|^2 < \rho\}$, and so these "balls" are very elliptical: of diameter of order magnitude ρ for the $\text{Im}\ \zeta_1$ axis, but of diameter of order of magnitude $\sqrt{\rho}$ in the other variables.

Similarly, the approach region is given essentially by

$$\{z: |1 - z_1| < \text{constant}\ \delta(z), \sum_{K=2}^{n} |z_k|^2 \leq \text{constant}\ \delta(z)\}$$

which is conical in the z_1 variable, but "parabolic" in the $z_2,\ldots,z_n$ variables. These observations will be of capital importance below.

Additional References for Chapter I

§§1 - 6. For detailed expositions of the facts about potential theory in domains in $\mathbb{R}^N$ with smooth boundaries see e.g. Privalov and Kuznetzov [15], Aronszajn and Smith [1], K. T. Smith [18], and K. O. Widman [25]. This theory is of course a generalization of the more familiar situation of the unit disc or the half-space in $\mathbb{R}^N$. The basic results in that context are in Zygmund [26], Stein and G. Weiss [22], and Stein [19].

§7. For general facts about the Bergman kernel (and Szegö kernel) see Bergman [2], and [3]; also Weil [24] and Helgason [5].

§8. The treatment of the unit ball in $\mathbb{C}^n$ as a bounded symmetric domain, Siegel [17], Hua [8], Pjateskii-Shapiro [14], Koranyi [10], [11].

Chapter II: Fatou's theorem

In this chapter we prove an analogue of Fatou's theorem for holomorphic functions in a bounded domain $\mathcal{D}$ with smooth boundary. It is to be emphasized that here assumptions of pseudo-convexity play no role. The main idea is to try to make estimates analogous to (8.14) for holomorphic functions while by-passing explicit information about the Poisson-Szegö kernel for $\mathcal{D}$ - knowledge about which we would, in any case, have little hope of getting.

§9. The first maximal inequality and its application

Let $\mathcal{D}$ be a bounded domain in $\mathbb{C}^n$ with smooth (C^2) boundary. For each $\zeta \in \partial\mathcal{D}$, let ν_ζ denote the unit outward normal at ζ. For each $\alpha > 0$, we define an approach region $\mathcal{A}_\alpha(\zeta)$, with vertex ζ, by the equation

$$\mathcal{A}_\alpha(\zeta) = \{z \in \mathcal{D} : |(z-\zeta)\cdot\bar{\nu}_\zeta| < (1+\alpha)\delta_\zeta(z), |z-\zeta|^2 < \alpha\delta_\zeta(z)\} .$$

Here $\delta_\zeta(z)$ = minimum of the distances from z to $\partial\mathcal{D}$ and from z to the tangent plane at ζ. Notice that when $\mathcal{D}$ is the unit ball $\delta(z) = \delta_\zeta(z)$ and $\mathcal{A}_\alpha(\zeta)$ is essentially the approach region $\mathcal{A}_\alpha(\zeta)$ defined in §8.

We shall say that F is <u>admissibly bounded at</u> ζ, $\zeta \in \partial\mathcal{D}$, if $\sup_{z \in \mathcal{A}_\alpha(\zeta)} |F(z)| < \infty$, for some $\alpha > 0$; F has an admissible limit at ζ if $\lim_{z \to \zeta, z \in \mathcal{A}_\alpha(\zeta)} F(z)$ exists, for all $\alpha > 0$.

We now consider two types of "balls" on $\partial\mathcal{D}$. For any $\rho > 0$, and $\zeta^0 \in \partial\mathcal{D}$, let

$$B_1(\zeta^0,\rho) = \{\zeta \in \partial\mathcal{D}: |\zeta-\zeta^0| < \rho\}$$

$$B_2(\zeta^0,\rho) = \{\zeta \in \partial\mathcal{D}: |(\zeta-\zeta^0)\cdot\overline{\nu}_{\zeta^0}| < \rho, |\zeta-\zeta^0|^2 < \rho\}$$

We shall see below that each of this family of balls satisfies the condition α) to δ) of §5. Thus we may consider the corresponding maximal functions f_j^*, for $j = 1,2$,

$$f_j^*(\zeta^0) = \sup_{\rho > 0} m(B_j(\zeta^0,\rho))^{-1} \int_{B_j(\zeta^0,\rho)} |f(\zeta)|d\sigma(\zeta)$$

where $d\sigma = dm =$ is the induced measure on $\partial\mathcal{D}$, and $f \in L^p(d\sigma)$. We also consider the superposition of these two maximal functions, $M(f)$ defined by

$$(Mf)(\zeta) = (f_1^*)_2^*(\zeta) .$$

Because of theorem 2 in §5 we can assert that:

$$\|M(f)\|_p \leq A_p\|f\|_p , \text{ if } f \in L^p(d\sigma) , \text{ for } 1 < p \leq \infty .$$

However, there is no L^1 inequality for M. The main estimate that will be made in this section is as follows:

<u>Lemma.</u> <u>Suppose</u> u <u>is continuous in</u> $\overline{\mathcal{D}}$ <u>and pluri-subharmonic in</u> $\mathcal{D}$. <u>Let</u> f <u>be the restriction of</u> u <u>to</u> $\partial\mathcal{D}$. <u>Then for each</u> α,

(9.1) $$\sup_{z \in \mathcal{a}_\alpha(\zeta)} |u(z)| \leq A_\alpha (Mf)(\zeta) .$$

We first clarify certain simple geometric notions arising from the interplay of the boundary $\partial\mathcal{D}$ and the complex structure of the ambient space $\mathbb{C}^n$. $\mathbb{C}^n$ has a standard hermitian inner product, whose real-part is the standard inner product of $\mathbb{R}^{2N}$, when $\mathbb{R}^{2N}$ is identical with $\mathbb{C}^n$ in the usual way, with $N = 2n$. With each point $\zeta \in \partial\mathcal{D}$ we have associated the unit outward normal. Let T_ζ denote the (real) tangent space at ζ. We can identify T_ζ (in the obvious way) as a 2n-1 dimensional real subspace of $\mathbb{C}^n$, which is the (real) orthogonal complement of $\{\mathbb{R}\nu_\zeta\}$. Similarly let C_ζ be the complex orthogonal complement of $\{\mathbb{C}\nu_\zeta\}$. We write N_ζ^o as the (real) subspace $\{\mathbb{R}i\nu_\zeta\}$, and N_ζ as the complex subspace $\{\mathbb{C}\nu_\zeta\}$. Then we have the orthogonal sum decompositions

$$(9.2) \qquad \begin{cases} \mathbb{C}^n = N_\zeta \oplus C_\zeta & (\text{over } \mathbb{C}) \\ T_\zeta = N_\zeta^o \oplus C_\zeta & (\text{over } \mathbb{R}) \end{cases}$$

Observe that T_ζ, C_ζ, N_ζ, and N_ζ^o have real dimensions respectively 2n-1, 2n-2, 2, and 1. Since $\partial\mathcal{D}$ was assumed to be of class C^2, the functions $\zeta \longrightarrow T_\zeta$, C_ζ, N_ζ etc. depend in a C^1 way on ζ. For each $\zeta \in \partial\mathcal{D}$ we can define the orthogonal projection π_ζ of $\mathbb{C}^n$ to N_ζ. Then $B_2(\zeta^o,\rho)$ can be written as

$$\{\zeta \in \partial\mathcal{D} : |\pi_{\zeta^o}(\zeta-\zeta^o)| < \rho, |\zeta-\zeta^o| < \rho^{\frac{1}{2}}\} .$$

Let us prove that the family B_2 satisfies the conditions (α) to (δ) of §5. Now (α) and (β) are obvious, and (δ) is an immediate consequence of the easily verified estimate $m[B_2(\zeta^o,\rho)] \sim c_o\rho^n$, as as $\rho \longrightarrow 0$. We turn therefore to (γ). We must show that if $u \in B_2(x,\rho) \cap (B_2(y,\rho)$, then there is a constant c , so that $B(y,\rho) \subset B(x,c\rho)$. Since $u \in B_2(x,\rho)$, $|x-u| < \rho^{\frac{1}{2}}$; similarly $|x-y| < \rho^{\frac{1}{2}}$; therefore $|x-y| < 2\rho^{\frac{1}{2}}$. Now let v be any point in $B(y,\rho)$; then $|v-y| < \rho^{\frac{1}{2}}$, and thus

$$(9.3) \qquad |x-v| < 3\rho^{\frac{1}{2}} .$$

Also,

$$\pi_x(x-v) = \pi_x(x-u) + \pi_y(u-v) + [\pi_x-\pi_y](u-v) .$$

Now the first term on the right side is less than ρ , since $u \in B_2(x,\rho)$; similarly the second term also does not exceed ρ . For the third term observe that $|(\pi_x-\pi_y)(u-v)| \leq |\pi_x-\pi_y\| |u-v| \leq \|\pi_x-\pi_y\| 2\rho^{\frac{1}{2}}$, since both u and v lie in $B_2(y,\rho)$. Also $\|\pi_x-\pi_y\| \leq c_1|x-y| \leq c_1 2\rho^{\frac{1}{2}}$. Altogether then,

$$(9.4) \qquad |\pi_x(x-v)| \leq (3+4c_1)\rho .$$

Combining (9.3) with (9.4) shows that $B_2(x,c\rho) \supset B_2(y,\rho)$ if $c = \max(9, 3+4c_1)$, proving that the family B_2 satisfies the required conditions. The proof that the family B_1 also satisfies these conditions is similar, and if anything is more standard, so we omit it. We can now turn to the proof of the lemma.

In proving (9.1) we may assume that z is sufficiently close to ζ , for otherwise the estimate is semi-trivial. After a possible translation and unitary linear transformation in the ambient space $\mathbb{C}^n$

we may assume that the boundary point $\zeta \in \partial\mathcal{D}$ is the origin, that the outward unit normal at 0 is in the negative y_1 direction, (here $z_1 = x_1 + iy_1$) and thus $N_\zeta = \{(z_1, 0, \ldots, 0)\}$, while $C_\zeta = \{(0, z_2, z_3 \ldots z_n)\}$.

Next for each fixed approach region $\mathcal{Q}_\alpha(0)$, we can find sufficiently small constant c_1, and another $\alpha' > \alpha$, so that if $z \in \mathcal{Q}_\alpha(0)$, and if $z' = (z_k')$ is such that $|z_1 - z_1'| < c_1 y_1$, $|z_k - z_k'|^2 < c_1 y_1$, $k = 2, \ldots, n$, then $z' \in \mathcal{Q}_{\alpha'}(0)$.

Now for z ranging over $\mathcal{Q}_{\alpha'}(0)$, denote by $\hat{z}$ the projection of z to $\partial\mathcal{D}$ along the y_1 direction; that is if $\hat{z} = (\hat{x}_1 + i\hat{y}_1, \ldots, \hat{x}_n + i\hat{y}_n)$, then $\hat{x}_k = x_k$, $k = 1, \ldots, n$, $\hat{y}_k = y_k$, $k = 2, \ldots, n$, and $\hat{z} \in \partial\mathcal{D}$. The mapping $z \longrightarrow \hat{z}$ is well-defined and smooth near the origin. Our preliminary estimate is the claim

$$(9.5) \qquad |u(z)| \leq c_\alpha f_1^*(\hat{z}) .$$

In fact since $|u(z)|$ is subharmonic it is clearly majorized by the Poisson integral of $|f|$ (in the sense of the first part of Chapter I). In addition the line segment joining z to $\hat{z}$ lies in a cone $\Gamma_\beta(\hat{z})$, for β sufficiently large, again assuming that z is sufficiently close to the origin. Therefore (9.5) follows theorem 3, part (b), in §5.

We now restrict z to $\mathcal{Q}_\alpha(0)$, and use the pluri-subharmonicity of $|u|$ to imply subharmonicity in the variables $z_1, \ldots, z_n$ separately. Thus

$$|u(z)| \leq \pi^{-n} \eta^{-n} \int_P |u(z_1 + \zeta_1, z_2 + \zeta_2, \ldots, z_n + \zeta_n)| d\omega(\zeta) .$$

Here P is the polydisc $|\zeta_1| < \eta$, $|\zeta_2|^2 < \eta, \ldots, |\zeta_n|^2 < \eta$, and $d\omega$ is the Euclidean element of volume.

Next with $y_1 > 0$ fixed, take $\eta = c_1 y_1$. Then as we have seen $z+\zeta \in \mathcal{A}_{\alpha'}(0)$, and estimate (9.5) may be used. Observe also that for all $y_1 > 0$, $(z+\zeta)^\wedge \in B_2(0, dy_1)$ for a fixed (sufficiently large) d. The result is therefore

$$(9.6) \qquad |u(z)| \le c\, y_1^{-n} \int_{B_2(0,dy_1)} |f_1^*(\zeta)|\,d\sigma(\zeta) .$$

Since, as we have already observed, $m(B_2(0,y_1)) \sim c_0\, y_1^n$, we get from (9.6) our desired conclusion (9.1) at any fixed point on the boundary ζ. But all the estimates we have made depend uniformly on ζ, as ζ ranges over $\partial\mathcal{D}$, and (9.1) is completely proved.

The main result of this section is the following generalization of Fatou's theorem.

Theorem 9. *Suppose $F(z)$ is holomorphic and bounded in a bounded domain $\mathcal{D}$ with C^2 boundary. Then F has admissible limits at almost every boundary point.*

The theorem is a consequence of a more general one dealing with the H^p classes. For every $0 < p < \infty$, define $H^p(\mathcal{D})$ to be of class $F(z)$ holomorphic in $\mathcal{D}$, for which

$$\sup_{\varepsilon > 0} \int_{\partial\mathcal{D}_\varepsilon} |F(z)|^p d\sigma_\varepsilon(z) < \infty$$

for any fixed family of approximating domains, as in §2. Since $|F(z)|^p$

is subharmonic this condition is independent of the particular approximating domains used, and is equivalent with the requirement that $|F(z)|^p$ has a harmonic majorant - where harmonic is taken in the standard sense of $\mathbb{R}^N$, $N = 2n$; for this see §4.

Theorem 10. Suppose F belongs to $H^p(\mathcal{D})$, $p > 0$, with $\mathcal{D}$ as above. Then F has admissible limits at almost every boundary point.

We now prove theorem 10. We let $\mathcal{D}_o$ be a subdomain of $\mathcal{D}$ of the type used in the proof of theorem 1 in §4. It will suffice to show that F has an admissible limit at almost every point of the common boundary of $\mathcal{D}$ and $\mathcal{D}_o$. Let ν be an outward unit normal at a fixed common boundary point, with the property that

$$\overline{\mathcal{D}}_o - \varepsilon\nu \subset \mathcal{D}, \text{ for sufficiently small positive } \varepsilon .$$

Our first step will be to show that for almost all $\zeta \in \partial\mathcal{D}_o$, $\lim_{\varepsilon \to 0} F(\zeta-\varepsilon\nu)$ exists (we call this limit $F(\zeta)$), and

(9.7) $$\int_{\partial\mathcal{D}_o} |F(\zeta-\varepsilon\nu) - F(\zeta)|^p d\sigma_o(\zeta) \longrightarrow 0 , \quad \text{as } \varepsilon \longrightarrow 0 .$$

Now let $s(z) = |F(z)|^{p/2}$. Because $\sup_{\varepsilon > 0} \int_{\partial\mathcal{D}_\varepsilon} (s(z))^2 d\sigma(z) < \infty$, we know there exists a harmonic function h , which is the Poisson integral of an L^2 function on $\partial\mathcal{D}$, so that

$$|F(z)|^{p/2} \leq h(z) .$$

By the use of the majorization in theorem 3, h is non-tangentially bounded almost everywhere on $\partial\mathcal{D}$, and so the same is true for $|F|$.

However every boundary point ζ^o is either (1) an interior point of $\mathcal{B}$, or (2) a boundary point of $\mathcal{B}$ also. Thus F is non-tangentially bounded at $\partial\mathcal{B}_o$ and has non-tangential limits at almost all points of $\partial\mathcal{B}_o$. In particular, $\lim_{\varepsilon \to o} F(\zeta-\varepsilon\nu) = F(\zeta)$ exists for almost every $\zeta \in \partial\mathcal{B}$. However by the maximal estimate just used $\sup_{\varepsilon > o} |F(\zeta-\varepsilon\nu)|^{p/2}$ is dominated by an L^2 function on $\partial\mathcal{B}_o$, and so (9.7) follows by the dominated convergence theorem.

For every pair of positive integers j and k, let $u(z) = |F(z - \nu/k) - F(z - \nu/j)|^{p/2}$. Then u is continuous in $\overline{\mathcal{B}}_o$ and is pluri-subharmonic in $\mathcal{B}_o$. Thus we can apply the lemma to u in $\mathcal{B}_o$ and the L^2 boundedness of M, to obtain

$$\int_{\partial\mathcal{B}_o} (\sup_{z \in \mathcal{A}_\alpha(\zeta)} u(z))^2 d\sigma_o(\zeta) \leq A \int_{\partial\mathcal{B}_o} |F(\zeta - \nu/k) - F(\zeta - \nu/j)|^p d\sigma_o(\zeta) .$$

Letting j tend to infinity, we get in view of (9.7),

$$\begin{aligned} \int_{\mathcal{B}_o} \sup_{z \in \mathcal{A}_\alpha(\zeta)} |F(z - \nu/k) - F(z)|^p d\sigma_o(\zeta) \leq \\ \leq A \int_{\mathcal{B}_o} |F(\zeta - \nu/k) - F(\zeta)|^p d\sigma_o(\zeta) . \end{aligned} \tag{9.8}$$

Now $F(z) = F(z - \nu/k) + \{F(z) - F(z - \nu/k)\}$. $F(z - \nu/k)$ has admissible limits at every point of $\partial\mathcal{B}_o$, for each k, since $F(z - \nu/k)$ is continuous in $\overline{\mathcal{B}}_o$, and

$$\int \sup_{z \in \mathcal{A}_\alpha(\zeta)} |F(z - \nu/k) - F(z)|^p d\sigma_o(\zeta)$$

tends to zero as $k \longrightarrow \infty$, because of (9.8) and (9.7). It is now a very routine argument to show that as a consequence F has admissible limits at almost every point of $\partial \mathcal{D}_o$.

The proof of theorem 10 also gives the following corollary:

__Corollary.__ _Suppose_ $F \in H^p(\mathcal{D})$, $0 < p < \infty$. _Then_

$$\int_{\partial \mathcal{D}} \sup_{z \in \mathcal{Q}_\alpha(\zeta)} |F(z)|^p d\sigma(\zeta) \leq A_{p,\alpha} \sup_{\varepsilon > o} \int_{\partial \mathcal{D}_\varepsilon} |F(\zeta)|^p d\sigma_\varepsilon(\zeta) .$$

10. The second maximal inequality and its application

Our aim is to extend the result guaranteeing the existence almost everywhere of admissible limits to the Nevanlinna class. The argument we gave above for H^p will have to be modified on two counts: First, we need a refined variant of the maximal function M , which applies to $L^1(\partial \mathcal{D})$ and more precisely, to finite measures. Secondly we shall have to exploit logarithmic pluri-subharmonicity to be able to do without an analogue of (9.7)

We begin by considering a continuum of maximal functions, each determined by a parameter λ , with $\lambda \geq 1$. The "balls" associated with this function are:

$$B^\lambda(\zeta^o,\rho) = \{\zeta \in \partial \mathcal{D} : |(\zeta - \zeta^o)\cdot \bar{\nu}_{\zeta^o}| < \lambda\rho , |\zeta - \zeta^o|^2 < \rho\} .$$

Notice that if $\lambda = 1$, these balls reduce to $B_2(\zeta^o,\rho)$, while when $\lambda \geq \rho^{-\frac{1}{2}}$ the balls $B^\lambda(\zeta^o,\rho)$ are equal to $B_1(\zeta^o,\rho^{\frac{1}{2}})$.

A crucial fact that we want to emphasize about these balls is that for each λ they satisfy the condition (α) to (δ) in §5, and they do so with bounds that can be taken to be independent of λ. The argument is merely a repetition of the one given for B_2. In fact (α) and (β) are obvious and (δ) is a consequence of the easily verified estimate

$$0 < b \leq \frac{m(B^\lambda(\zeta^o,\rho))}{\lambda\rho^n} \leq a \ , \quad \text{when} \quad \rho \leq \lambda^{-2} \ ;$$

also recall that $B^\lambda(\zeta^o,\rho)$ are effectively $B_1(\zeta^o,\rho^{\frac{1}{2}})$ when $\rho > \lambda^{-2}$.

It remains to prove (γ). As before

$$(9.3) \qquad |x-v| < 3\rho^{\frac{1}{2}} \ .$$

Also

$$\pi_x(x-v) = \pi_x(x-u) + \pi_y(u-v) + (\pi_x - \pi_y)(u-v) \ .$$

The norm of the first term on the right side does not exceed $\lambda\rho$, since $u \in B^\lambda(x,\rho)$. Similarly for the second term we have the estimate $2\lambda\rho$, since both u and $v \in B^\lambda(y,\rho)$. The estimate for the third term is, as before, $4c_1\rho$. Altogether then

$$(9.4)' \qquad |\pi_x(x-v)| \leq (3\lambda + 4c_1)\rho \ .$$

This with (9.3) shows that $B^\lambda(x,c\rho) \supset B^\lambda(y,\rho)$ if

$c = \max(9, \frac{3\lambda + 4c_1}{\lambda}) \leq \max(9, 3 + 4c_1)$.

As a result, if we define f^*_λ by

$$f^*_\lambda(\zeta^o) = \sup_{\rho > o} m(B^\lambda(\zeta^o,\rho))^{-1} \int_{B^\lambda(\zeta^o,\rho)} |f(\zeta)| d\sigma(\zeta)$$

and $f \in L^1(d\sigma)$, then

(10.4) $$m\{\zeta: f^*_\lambda(\zeta) > \alpha\} \leq K/\alpha \int_{\partial\mathcal{D}} |f(\zeta)|d\sigma(\zeta)\ , \quad \text{all} \quad \alpha > 0\ ,$$

where K does not depend on λ. We can similarly define the maximal functions associated with general finite Borel measures $d\mu(\zeta)$ on $\partial\mathcal{D}$, instead of merely absolutely continuous ones like $f(\zeta)d\sigma(\zeta)$. We write

$$f^*_\lambda(d\mu)(\zeta^o) = \sup_{\rho > o} m(B^\lambda(\zeta^o,\rho))^{-1} \int_{B^\lambda(\zeta^o,\rho)} |d\mu(\zeta)|\ .$$

Then again we have

(10.4)' $$m\{\zeta: f^*_\lambda(d\mu)(\zeta) > \alpha\} \leq K/\alpha \int_{\partial\mathcal{D}} |d\mu(\zeta)|\ , \quad \text{all} \quad \alpha > 0\ .$$

This can be proved as an easy consequence of (10.4) by approximating $d\mu$ by a sequence of L^1 functions where L^1 norm does not exceed the total variation of $d\mu$; or by deducing it from the covering lemma for the family B^λ in the same way as (9.4).

We are now in a position to define the controlling maximal function:

(10.5) $$\tilde{M}(d\mu)(\zeta^o) = f^*(d\mu)(\zeta^o) + \sum_{k=1}^{\infty} 2^{-k} f^*_{2^k}(d\mu)(\zeta^o)\ ,$$

where f^* (without subscript) is the maximal function taken with respect to the genuine balls $B_1(\cdot,\rho)$. The properties we shall need are as follows.

Lemma 1.

(a) For each finite measure $d\mu$, $\tilde{M}(d\mu)(\zeta^o)$ is finite for almost every $\zeta^o \in \partial\mathcal{D}$, and in particular,

(10.6) $$m\{\zeta \in \partial\mathcal{D}: \tilde{M}(d\mu)(\zeta) > \gamma\} \leq K'/\gamma \int_{\partial\mathcal{D}} |d\mu(\zeta)|\ .$$

(b) <u>Suppose</u> u <u>is continuous in</u> $\overline{\mathcal{D}}$ <u>and is pluri-subharmonic in</u> $\mathcal{D}$. <u>Let</u> f <u>be its restriction to</u> $\partial\mathcal{D}$. <u>Then for every</u> $\alpha > 0$,

$$(10.7) \qquad \sup_{z \in \mathcal{A}_\alpha(\zeta)} |u(z)| \leq A_\alpha(\tilde{M}f)(\zeta) .$$

To prove (a) we begin by observing that $\sum_{k=o}^{\infty} 2^{-k/2} < 3$. Now if $\sum_{k=o}^{\infty} 2^{-k} f^*_{2^k}(d\mu)(\zeta^o) = \sum_{k=o}^{\infty} 2^{-k/2}\, 2^{-k/2} f^*_{2^k}(d\mu)(\zeta^o) > \gamma$, then for at least one k , $2^{-k/2} f^*_{2^k}(d\mu)(\zeta^o) > \gamma/3$. Therefore

$\{\zeta\colon \tilde{M}(d\mu) > \gamma\} \subset \bigcup_k \{\zeta\colon 2^{-k/2} f^*_{2^k}(d\mu)\zeta > \gamma/3\}$ and

$m\{\zeta\colon \tilde{M}(d\mu) > \gamma\} \leq \sum_{k=o}^{\infty} \frac{3K2^{-k/2}}{\gamma} \int_{\partial\mathcal{D}} |d\mu| \leq 9K/\alpha \int_{\partial\mathcal{D}} |d\mu|$, which is (10.6).

To prove (10.7) we need first introduce some further definitions. For the family of balls B_1, B_2, and B^λ we consider the associated mean-values. Thus we write

$$\mathcal{M}_j(\rho{:}f)(\zeta^o) = m[B_j(\zeta^o,\rho)]^{-1} \int_{B_j(\zeta^o,\rho)} |f(\zeta)|\, d\sigma(\zeta) , \qquad j = 1,2 .$$

Also

$$\mathcal{M}^\lambda(\rho{:}f)(\zeta^o) = m(B^\lambda(\zeta^o,\rho))^{-1} \int_{B^\lambda(\zeta^o,\rho)} |f(\zeta)|\, d\sigma(\zeta) .$$

Clearly $f^*_j(\zeta^o) = \sup_{o<\rho} \mathcal{M}_j(\rho{:}f)(\zeta^o)$, $j = 1,2$ and

$$f^*_\lambda(\zeta^o) = \sup_{o<\rho} \mathcal{M}^\lambda(\rho{:}f)(\zeta^o) , \quad \lambda \geq 1 .$$

We shall require the following facts about the superposition of the mean-values:

(10.8) $$\mathcal{M}_2(\rho:\mathcal{M}_1(\lambda\rho:f))(\zeta) \le c\,\mathcal{M}^{\lambda}(c\rho:f)(\zeta) ,$$

for an appropriate constant C independent of ρ and λ, if $\lambda\rho \le \rho^{\frac{1}{2}}$.

(10.9) $$\mathcal{M}_2(\rho:\mathcal{M}_1(\lambda\rho:f))(\zeta) \le c\,\mathcal{M}_1(c\lambda\rho:f)(\zeta) \quad \text{if} \quad \lambda\rho \ge \rho^{\frac{1}{2}} .$$

To prove (10.8) we let $\Psi_{\lambda\rho}(u,v)$ be the characteristic functions of the set where $u \in B_1(v,\lambda\rho)$; and $\chi_\rho(w,u)$ the characteristic function of the set where $u \in B_2(w,\rho)$. We need to consider

(10.10) $$\int_{\partial\mathcal{D}} \chi_\rho(w,u)\Psi_{\lambda\rho}(u,v)d\sigma(u) .$$

For what pair w and v is the integral non-zero? Clearly what is required is the existence of a u, such that $|u-v| < \lambda\rho$; also $|w-u| < \rho^{\frac{1}{2}}$ and $|\pi_w(w-u)| < \rho$. We conclude therefore that $|w-u| < \rho^{\frac{1}{2}} + \lambda\rho < 2\rho^{\frac{1}{2}}$, and also

$$|\pi_w(w-v)| \le |\pi_w(w-u)| + |\pi_w(u-v)| < \rho + \lambda\rho < 2\lambda\rho .$$

Hence if the integral (10.10) does not vanish then $v \in B_\lambda(c\rho,w)$ where $c = 4$. Now the value of the integral is less than the measure of the set $\{u \in \partial\mathcal{D} : |u-v| < \lambda\rho, |\pi_w(w-u)| < \rho\}$ which is less than a constant $\times\ \rho\cdot(\lambda\rho)^{2n-2}$. Hence if $\eta(w,v)$ is the characteristic function of $\{v \in B_\lambda(c\rho,w)\}$, then (10.10) is less than

$$\text{constant} \times \lambda^{2n-2}\cdot\rho^{2n-1}\cdot\eta(w,v) .$$

If we divide through by the normalizing factors $m(B_1(\cdot,\lambda\rho))$, $m(B_2(\cdot,\rho))$

and $m(B^{\lambda}(\cdot, c\rho))$, which have orders of magnitudes $(\lambda\rho)^{2n-1}, \rho^n$, and $\lambda\rho^n$ respectively, we then get (10.8). The proof of (10.9) is similar and may be left to the reader.

To complete the proof of part (b) of Lemma 1, we return to the analysis used to prove the lemma in §9, and in particular giving (9.5) and (9.6).

Recall that the boundary point is the origin, and the outward normal direction is the negative y_1 direction (with $z = (z_1, \ldots, z_n)$ and $z_1 = x_1 + iy_1$) . $\alpha > 0$ is fixed and c_1 and α' have been chosen so that if $\alpha \in \mathcal{A}_\alpha(0)$ and $|z_1 - z_1'| < c_1 y_1$, $|z_k - z_k'|^2 < c_1 y_1$, $k = 2, \ldots, n$, then $z' \in \mathcal{A}_{\alpha'}(0)$. For any $z \in \mathcal{D}$, sufficiently close to 0 , we let $\hat{z}$ denote its perpendicular projection (along y_1 direction) on $\partial \mathcal{D}$. Instead of (9.5), which is a consequence of part (b) of theorem 3 in §5, we use the more precise part (a) of that theorem. We get, therefore, in the notation introduced above, for $z \in \mathcal{A}_{\alpha'}(0)$,

$$(10.11) \qquad |u(z)| \leq c_{\alpha'}' \sum_{k=1}^{\infty} 2^{-k} \, \mathcal{M}_1(2^k y_1 : f)(\hat{z}) \; .$$

We now restrict z to $\mathcal{A}_\alpha(0)$, and let $P(z)$ be the poly-disc $P(z) = \{z' : |z_1 - z_1'| < c_1 y_1, |z_k - z_k'|^2 < c_1 y_1, \; k = 2, \ldots, n\}$ and use the fact that

$$(10.12) \qquad |u(z)| \leq \frac{1}{|P(z)|} \int_{P(z)} |u(z')| \, d\omega(z') \; ,$$

where $|P(z')|$ denotes the measure of $P(z)$.

In applying (10.12) to (10.11) it is useful to note that if $\rho_2 \leq \rho_1 \leq c\rho_2$, for some fixed constant c, $c > 1$, then

$$\mathcal{M}_j(\rho_2:f) \leq C\,\mathcal{M}_j(\rho_1:f) , \tag{10.13}$$

with C independent of f, ρ_1 and ρ_2. Using this in (10.11) together with (10.12) we get, after some easy simplifications,

$$|u(z)| \leq A_\alpha \sum_{k=1}^{\infty} 2^{-k}\mathcal{M}_2(y_1 : \mathcal{M}_1(2^k y_1 : f))(0) .$$

We can now invoke (10.8) and (10.9). The result is

$$|u(z)| \leq A'_\alpha \left\{ \sum_{2 \leq 2^k < y_1^{-\frac{1}{2}}} 2^{-k}\,\mathcal{M}^{2^k}(cy_1 : f)(0) + \sum_{2^k \geq y_1^{-\frac{1}{2}}} 2^{-k}\mathcal{M}_1(c2^k y_1 : f)(0) \right\}.$$

The first sum is clearly dominated by a constant multiple of

$$\sum_{k=1}^{\infty} 2^{-k} f^*_{2^k}(0)$$

since $\mathcal{M}^{2^k}(\rho:f)(\zeta) \leq f^*_{2^k}(\zeta)$. Similarly the second sum is dominated by

$$\left(\sum_{k=1}^{\infty} 2^{-k}\right) f^*(0) ,$$

and then altogether

$$\sup_{z \in \mathcal{a}_\alpha(0)} |u(z)| \leq A_\alpha \tilde{M}(f)(0) .$$

Since $\zeta = 0$ is a general point of the boundary we have therefore obtained $\sup_{z \in \mathcal{U}_\alpha(\zeta)} |u(z)| \le A_\alpha(\tilde{M}f)(\zeta)$, and the proof of Lemma 1 is complete.

For future reference we record some further inequalities dealing with the superposition of mean-values, whose proofs are very analogous to the one we gave for (10.8). Here $c > 1$ is a constant (independent of $\rho_1, \rho_2, \lambda_1 \dots$ etc.); but c need not be the same on different occasions.

$$(10.14)\quad \begin{cases} \mathcal{M}_j(\rho_1 : \mathcal{M}_j(\rho_2 : f)) \le c\, \mathcal{M}_j(c(\rho_1 + \rho_2) : f), \quad j = 1,2 . \\ \mathcal{M}_1(\rho_1 : \mathcal{M}_2(\rho_2, f)) \le c\, \mathcal{M}_2(c\rho_2 : \mathcal{M}_1(c\rho_1 : f)) . \end{cases}$$

Armed with lemma 1 and the inequalities (10.14) we can finally come to the analogue of Fatou's theorem for the Nevanlinna class.

We shall say that $F \in N(\mathcal{D})$, if F is holomorphic in $\mathcal{D}$ and

$$(10.15)\qquad \sup_{\varepsilon > 0} \int_{\partial \mathcal{D}_\varepsilon} \log^+ |F(z)|\, d\sigma_\varepsilon(z) < \infty .$$

Here $\log^+ u = \max(0, \log u)$.

Since $\log^+|F(z)|$ is subharmonic it follows from the results of §4 that (10.15) is equivalent with the statement that $\log^+|F(z)|$ has a harmonic majorant, and more precisely that there exists a positive, finite, measure $d\mu$ on $\partial \mathcal{D}$, so that if $h(z)$ is its Poisson integral, $h(z) = \int_{\partial \mathcal{D}} P(z,\zeta)\, d\mu(\zeta)$, then

(10.16) $$\log^+|F(z)| \leq h(z) .$$

Of course we assume that $\partial\mathcal{D}$ is of class C^2 as above.

Our result is

<u>Theorem 11. Suppose F belongs to $N(\mathcal{D})$. Then F has admissible limits at almost every boundary point.</u>

This theorem obviously extends theorems 9 and 10 of §8.

<u>Proof.</u> Let $\mathcal{D}_o$ be the sub-domain of $\mathcal{D}$ already used in the proof of theorem 10. $\mathcal{D}_o$ has part of its boundary in common with $\mathcal{D}$, and for a unit normal ν at a point of their common boundaries, we have $\overline{\mathcal{D}}_o - \varepsilon\nu \subset \mathcal{D}$, for all positive and sufficiently small ε. It suffices to prove that F has an admissible limit of almost every point of $\partial\mathcal{D}_o$.

Now consider

(10.17) $$\sup_{z \in \mathcal{A}_\alpha(\zeta)} \log|F(z-\nu/k) - F(z-\nu/j)| , \quad \text{for } \zeta \in \partial\mathcal{D}_o .$$

We shall show that the quantity in (10.17) tends to $-\infty$, as k and $j \longrightarrow \infty$, for almost every $\zeta \in \partial\mathcal{D}_o$. For those $\zeta \in \partial\mathcal{D}_o$, at which this happens for all α, it is a routine matter to show that F has an admissible limit at ζ. Now let $P(z,\zeta)$ be the Poisson kernel for $\mathcal{D}_o$. Then since $\log|F(z-\nu/k) - F(z-\nu/j)|$ is subharmonic, it follows that

$$\log|F(z-\nu/k) - F(z-\nu/j)| \leq \int_{\partial\mathcal{D}_o} P(z,\zeta)\log|F(\zeta-\nu/k) - F(\zeta-\nu/j)|d\sigma(\zeta)$$

$$\leq \int_{\partial\mathcal{D}_o} P(z,\zeta)\log^-|F(\zeta-\nu/k) - F(\zeta-\nu/j)|d\sigma(\zeta) + \int_{\partial\mathcal{D}_o} P(z,\zeta)\log^+|F(\zeta-\nu/k)|d\sigma(\zeta)$$

$$+ \int_{\partial\mathcal{D}_o} P(z,\zeta)\log^+|F(\zeta-\nu/j)|d\sigma(\zeta) .$$

We now consider a boundary point ζ^{0} of $\partial\mathcal{D}_{0}$. For convenience of calculation we shall assume, as we have already done previously, that the coordinates have been chosen in $\mathbb{C}^{n}$ so that $\zeta^{0} = (0)$, and the unit normal ν_{ζ^0} is given by $(i,0,\ldots,0)$. As in the discussion immediately following (10.11), for any $z \in \mathcal{A}_{\alpha}(\zeta^{0})$, let $P(z)$ denote the polydisc

$$P(z) = \{z': |z_1 - z_1'| < c_1y_1, |z_k - z_k'|^2 < c_1y_1,\ k = 2,\ldots,n\},$$

and recall that if c_1 is sufficiently small, the $z \in \mathcal{A}_{\alpha}(\zeta^{0})$ implies $P(z) \subset \mathcal{A}_{\alpha'}(\zeta^{0})$, for some $\alpha' > \alpha$.

By the pluri-subharmonicity of the logarithm of the absolute value of a holomorphic function we have

$$\log|F(z-\nu/k) - F(z-\nu/j)| \le \frac{1}{|P(z)|}\int_{P(z)} \log|F(z'-\nu/k) - F(z'-\nu/j)|d\omega(z').$$

We now define $\tilde{P}(z,\zeta)$ by

$$\tilde{P}(z,\zeta) = \frac{1}{|P(z)|}\int_{P(z)} P(z',\zeta)d\omega(z').$$

Then the above inequalities give, as bound for $\log|F(z-\nu/k) - F(z-\nu/j)|$,

$$\text{(10.18)}\qquad \int_{\partial\mathcal{D}_0} \tilde{P}(z,\zeta)\log^{-}|F(\zeta-\nu/k) - F(\zeta-\nu/j)|d\sigma(\zeta) + H_{1/k}(z) + H_{1/j}(z)$$

where

$$\text{(10.19)}\qquad H_{\varepsilon}(z) = \int_{\partial\mathcal{D}_0} \tilde{P}(z,\zeta)\log^{+}|F(\zeta - \varepsilon\nu)|d\sigma(\zeta).$$

We shall show that, in an appropriate sense, the last two terms of (10.18) remain bounded, while the first one goes to $-\infty$, as $k,j \to \infty$.

These last two terms are controlled by the following lemma, and lemma 1.

Lemma 2. There exists a finite measure $d\mu$ on $\partial\mathcal{D}_o$, so that

$$\sup_{\varepsilon > 0} \sup_{z \in \mathcal{a}_\alpha(\zeta^o)} |H_\varepsilon(z)| \leq A_\alpha \tilde{M}(d\mu)(\zeta^o) .$$

Let $u_\varepsilon(z) = \log^+|F(z-\varepsilon\nu)|$. Then $u_\varepsilon(z)$ is continuous in $\overline{\mathcal{D}}_o$ and pluri-subharmonic and so we can apply lemma 1 and, more particularly, the inequality following (10.13). This gives

$$|u_\varepsilon(z)| \leq A_\alpha \sum_{k=1}^{\infty} 2^{-k} \mathcal{m}_2(y_1: \mathcal{m}_1(2^k y_1:f_\varepsilon)(\zeta) ,$$

if $z \in \mathcal{a}_\alpha(\zeta)$, where $f_\varepsilon = u_\varepsilon\big|_{\partial\mathcal{D}_o}$. Since

$\sup_{\varepsilon > 0} \int_{\partial\mathcal{D}_o} |f_\varepsilon(\zeta)|d\sigma(\zeta) = \sup_{\varepsilon > 0} \int_{\partial\mathcal{D}_o} \log^+|F(\zeta-\varepsilon\nu)|d\sigma(\zeta) < \infty$, (see the corollary in §4, and the lemma in §3), we can find a measure $d\mu$ on $\partial\mathcal{D}_o$ and a subsequence $\{f_{\varepsilon'_k}\}$, so that $f_{\varepsilon'_k} \longrightarrow d\mu$ weakly. The The passage to the limit then gives

$$\log^+|F(z)| \leq A_\alpha \sum_{k=1}^{\infty} 2^{-k} \mathcal{m}_2(y_1: \mathcal{m}_1(2^k y_1:d\mu))(\zeta)$$

for $z \in \mathcal{a}_\alpha(\zeta)$. In particular, if $z = \zeta - \nu\varepsilon$, $\zeta \in \partial\mathcal{D}$ then $\varepsilon c_1 < y_1 \leq c_2\varepsilon$, and so by (10.13),

$$\log^+|F(\zeta-\nu\varepsilon)| \leq A_\alpha \sum_{k=1}^{\infty} 2^{-k} \mathcal{m}_2[c\varepsilon: \mathcal{m}_1(2^k c\varepsilon:d\mu)](\zeta), \quad \zeta \in \partial\mathcal{D}_o \tag{10.20}$$

The argument following (10.13) also shows that

$$\int_{\partial\mathcal{D}_o} \tilde{P}(z,\zeta)\varphi(\zeta)d\sigma(\zeta) \leq A_\alpha \sum_{j=1}^{\infty} 2^{-j} \mathcal{m}_2(y_1: \mathcal{m}_1(2^j y_1:\varphi))(\zeta^o) \tag{10.21}$$

for any continuous function φ on the boundary $\partial\mathcal{D}$. Now let $\varphi(\zeta) = \log^+|F(\zeta-\nu\varepsilon)|$, and substitute (10.20) into (10.21). We can simplify the resulting expression by the use of the inequalities (10.14). After a long but straightforward reduction, we get as in the argument following (10.11) that

$$\sup_{\varepsilon > 0} \sup_{z \in \mathcal{A}_\alpha(\zeta^0)} H_\varepsilon(z) \leq A_\alpha \tilde{M}(d\mu)(\zeta^0) , \qquad \zeta^0 \in \partial\mathcal{D}_0 ,$$

which is the statement of lemma 2.

By comparing the Poisson kernel $P(z,y)$ with the known Poisson kernel of a ball tangent to $\mathcal{D}_0$ at ζ and lying in $\mathcal{D}_0$, we get the estimate

$$P(z,\zeta) \geq c \frac{\delta(z)}{|z-\zeta|^{2n}} , \tag{10.22}$$

where $\delta(z)$ is the distance of z from the boundary. Thus since $\tilde{P}(z,\zeta) = \frac{1}{|P(z)|} \int_{P(z)} P(z',\zeta)d\omega(z')$, we obtain easily that

$$\tilde{P}(z,\zeta) \geq c_\alpha \delta^{-n} \chi(\zeta,\zeta^0) \tag{10.23}$$

if $z \in \mathcal{A}_\alpha(\zeta^0)$, and χ is the characteristic function of the "ball" $\zeta \in B_2(\zeta^0,\delta)$, and $\delta = \delta(z)$.

Now the potential theory of $\mathbb{R}^N$ shows that if F is in the Nevanlinna class, $\lim_{k\to\infty} F(\zeta - \nu/k)$ exists for almost every $\zeta \in \partial\mathcal{D}_0$. In fact the non-tangential limits exist at almost every point of $\mathcal{D}_0$ for F since $\log^+|F|$ has a harmonic majorant, which shows that F is non-tangentially bounded at almost all points of the boundary, (by theorem 3 in §5 and the corollary in §4). The local Fatou theorem (theorem 5 of §6) is then applicable, which shows that F has non-tangential limits at almost all points of $\mathcal{D}_0$, and this implies that $\lim_{k\to\infty} F(\zeta - \nu/k)$ exists for those $\zeta \in \partial\mathcal{D}$. Thus the non-positive functions $\log^-|F(\zeta-\nu/k) - F(\zeta-\nu/j)|$ tend to $-\infty$ almost everywhere in $\partial\mathcal{D}_0$, as $k,j \longrightarrow \infty$. Hence for every $\eta > 0$, there is a subset $E_\eta \subset \partial\mathcal{D}_0$, so that $m({}^cE_\eta) < \eta$, and $\log^-|F(\zeta-\nu/k) - F(\zeta-\nu/j)| \longrightarrow -\infty$ uniformly for $\zeta \in E_\eta$. In view of (10.23) we get that

$$(10.24) \qquad \sup_{z\in\mathcal{A}_\alpha(\zeta^0)} \int \tilde{P}(z,\zeta)\log^-|F(\zeta-\nu/k) - F(\zeta-\nu/j)|d\sigma(\zeta) \longrightarrow -\infty$$

as $k,j \longrightarrow \infty$, for every point ζ^0 which is a point of density of the set E_η (for some η) and hence for almost all points of $\partial\mathcal{D}_0$. (We say here that ζ^0 is a point of density of the set E_η, if

$$\frac{m(B_2(\zeta^0,\rho) \cap E_\eta)}{m(B_2(\zeta^0,\rho))} \longrightarrow 1\,, \text{ as } \rho \longrightarrow 0\,.)$$

When we combine (10.24) with lemma 2 via (10.18) we get that for almost all $\zeta^0 \in \partial\mathcal{D}_0$

$$\sup_{z\in\mathcal{A}_\alpha(\zeta^0)} \log|F(z-\nu/k) - F(z-\nu/j)| \longrightarrow -\infty\,, \text{ as } k,j \longrightarrow \infty\,.$$

For these ζ^0, $F(z)$ has a limit as $z \longrightarrow \zeta^0$, $z \in \mathcal{A}_\alpha(\zeta^0)$, and thus the existence of admissible limits almost everywhere is proved.

References for Chapter II

These results were announced in Stein [20]. That the maximal function with respect to skew balls $B_2(\cdot,\rho)$ or their analogues, plays a role in the boundary behavior of holomorphic functions was known previously. It was found independently by Hörmander [6] in the context of strictly pseudo-convex domains, and Stein [21], Koranyi [10], and Stein and N. Weiss [23], in the context of bounded symmetric domains. However the present situation seems to be substantially different from the previous cases, since one cannot take advantage of either strict pseudo-convexity or the explicit formulaes in the bounded symmetric case. This is basically the reason one needs to deal with the complicated interplay of the balls B_2 with the "classical" balls B_1. See also the research announcements of Malliavin [12] and [13], which deal with the cases of strictly pseudo-convex domains and product domains, by using Brownian motion.

Chapter III. Potential theory for strictly pseudo-convex domains

In this chapter we shall assume that $\mathcal{D}$ is a smooth (C^2) subdomain of $\mathbf{C}^n$ which is strictly pseudo-convex (see the definition below). The assumption of strict pseudo-convexity will allow us to introduce an appropriate metric in $\mathcal{D}$, and it is the potential theory of the Laplace-Beltrami operator of that metric which is the basic tool in the proof of theorem 12 below.

11. Potential theory in the context of a preferred Kahlerian metric

Recall (see §7) that for each $\zeta \in \partial\mathcal{D}$ we have considered the splitting of $\mathbf{C}^n$ given by $\mathbf{C}^n = N_\zeta \oplus C_\zeta$, and the induced splitting of the tangent plane $T_\zeta = N_\zeta^o \oplus C_\zeta$. Let now $\lambda(z)$ be any characterizing function for the domain $\mathcal{D}$ (see §2). We shall say that $\mathcal{D}$ is pseudo-convex if for each ζ the Hermitian quadratic form

$\mathcal{L}(t,t) = \sum_{i,j} \left(\frac{\partial\lambda}{\partial z_i \partial \bar{z}_j}\right) t_i \bar{t}_j$, restricted to the n-1 dimensional complex subspace C_ζ, is semi-definite. $\mathcal{D}$ is strictly (or strongly) pseudo-convex if $\mathcal{L}(t,t)$ on C_ζ is strictly positive definite. The following facts are well known; the first two are straightforward but the third is fundamental. (1) The definitions do not really depend on the particular characterizing function λ used for $\mathcal{D}$. (2) These notions are invariant under biholomorphic mappings which are smooth on the boundary. (3) If $\mathcal{D}$ has a smooth boundary then $\mathcal{D}$ is pseudo-convex if and only if $\mathcal{D}$ is a domain of holomorphy. (See Gunning and Rossi [4], and Hörmander [7], chapter 4).

For each $z \in \mathcal{D}$, sufficiently close to $\partial\mathcal{D}$, let $n(z) = \zeta$ be its normal projection on $\partial\mathcal{D}$: ζ is defined as the point on $\partial\mathcal{D}$ of minimum distance from z. The mapping $z \longrightarrow n(z)$ is at least of class C^1. For the projected boundary point $n(z)$ we have the splittings $\mathbb{C}^n = N_\zeta \oplus C_\zeta$, etc. just alluded to. If we write $N_z = N_{n(z)}$, $C_z = C_{n(z)}$, etc., we have extended these splittings to a neighborhood of $\partial\mathcal{D}$ in $\mathcal{D}$. Our principal definition is then as follows.

Definition. A metric $ds^2 = \Sigma\, g_{ij}(z)dz_i d\bar{z}_j$ defined on $\mathcal{D}$ will be called a preferred metric if it satisfies the following properties:

(1) the $g_{ij}(z)$ are continuous in $\mathcal{D}$

(2) the metric is Kahlerian

(3) $\Sigma\, g_{ij}(z)w_i\bar{w}_j \approx (\delta(z))^{-2}|w|^2$, for $w \in N_z$

(4) $\Sigma\, g_{ij}(z)w_i\bar{w}_j \approx (\delta(z))^{-1}|w|^2$, for $w \in C_z$

(5) $|\Sigma\, g_{ij}(z)w_i\bar{w}'_j| \leq c\delta(z)^{-1}|w||w'|$, for $w \in N_z$, $w' \in C_z$.

Here $\delta(z)$ denotes the distance of z from $\partial\mathcal{D}$. We write $A \approx B$ if the ration $|A|/|B|$ is bounded between two positive constants.

As the reader might guess, the inspiration for this definition comes after a careful scrutiny of the invariant (i.e. Bergman) metric for the unit ball, given by (8.4). The main fact about such metrics is their existence.

Lemma 1. *Suppose $\mathcal{D}$ has a C^2 boundary and is strictly pseudo-convex. Then $\mathcal{D}$ has a preferred metric.*

In effect, such a metric can be defined near $\partial\mathcal{D}$ by $g_{ij}(z) = \dfrac{\partial}{\partial z_i}\dfrac{\partial}{\partial \bar{z}_j}(\log 1/\delta(z))$. It is technically desirable, however, to modify this definition slightly by replacing the characterizing function $-\delta(z)$, by another $\lambda(z)$, which we can be certain is of class C^2; and by changing the definition for points strictly in the interior of $\mathcal{D}$. Thus let φ be a C^2 function which is 1 near $\partial\mathcal{D}$ and vanishes away from a small neighborhood of $\partial\mathcal{D}$. For a sufficiently large C write $G(z) = C|z|^2 + \varphi(z)\log(-1/\lambda(z))$, and

$$(11.1) \qquad g_{ij}(z) = \frac{\partial}{\partial z_i \partial \bar{z}_j}(G(z)) .$$

Property (1) is obvious. Property (2) follows immediately from (11.1): the Kahler conditions $\frac{\partial g_{ij}}{\partial \bar{z}_k} = \frac{\partial g_{ik}}{\partial \bar{z}_j}$ hold (at least in the sense of distributions, which suffices for our purposes below)*.

In proving the properties (3), (4) and (5) it suffices to take z sufficiently close to $\partial\mathcal{D}$, because the case when z lies strictly in the interior of $\mathcal{D}$ is taken care of by choosing the constant C (in the definition of $G(z)$) to be large and fixed.

Upon differentiation, we get

$$(11.2) \qquad g_{ij}(z) = \lambda^{-2}\{\lambda_{z_i}\lambda_{\bar{z}_j} - \lambda\lambda_{z_i\bar{z}_j}\} .$$

As an aid to the calculations, we introduce for each z near $\partial\mathcal{D}$ a coordinate system compatible with the splitting $\mathbb{C}^n = N_z \oplus C_z$. We take $z = (z_1,\ldots,z_n)$ so that $N_z = \{(z_1,0,\ldots,0)\}$, and $C_z = \{(0,z_2,z_3,\ldots,z_n)\}$, and $n(z) = (0,\ldots,)$. We also observe then that $\delta(z) = y_1 + O(\delta^2)$, that $-\lambda(z) \approx \delta(z)$, and more precisely that

---

*Footnote: See also the footnote following (11.5).

$-\lambda(z) = ay_1 + O(\delta^2)$, where $a > 0$ for $z \longrightarrow \zeta \in \partial \mathcal{D}$, with $a = a(\zeta) = \lim_{a \to \zeta} \frac{-\lambda(z)}{\delta(z)} = |\nabla\lambda(\zeta)|$. Also $|\frac{\partial\lambda}{\partial z_1}|^2 = (1/4)a^2$. We write for Q_2 the matrix $\{\lambda_{z_i \bar{z}_j}\}$, restricted to C_z . Notice that $Q_2(z)$ is strictly positive definite for z near $\partial \mathcal{D}$ by our assumption of strict pseudo-convexity. The matrix of ζ_{ij} given by (11.2) can then be written as follows.

$$g_{ij} = \lambda^{-2} \cdot \left(\begin{array}{c|c} a^2/4 + O(\lambda) & O(\lambda) \\ \hline O(\lambda) & -\lambda Q_2 + O(\lambda^2) \end{array} \right) \tag{11.3}$$

From this, the other properties of the metric are obvious, and the lemma is proved.

By Cramer's rule the determinant of the matrix g_{ij} is easily seen to be $a^2/4 \cdot (-\lambda)^{-n-1} \det Q_2 + O(\lambda^{-n})$. Its inverse matrix is then given, again in view of Cramer's rule, by

$$g^{ij} = \lambda^2 \left(\begin{array}{c|c} 4/a^2 + O(\lambda) & O(1) \\ \hline O(1) & (-\lambda)^{-1} Q_2^{-1} + O(1) \end{array} \right) \tag{11.4}$$

Our objective will be to apply the potential theory of the Laplace-Beltrami operator of the metric (11.3) which we know can be written as

$$\Delta f = 4 \sum g^{ij} \frac{\partial f}{\partial \bar{z}_i \partial z_j} \ . \tag{11.5}$$ (*)

(*) Strictly speaking, the usual definition of the Laplace-Beltrami operator, and the calculation leading to (11.5) require that the metric is of class C^1 . However the reader may easily convince himself that as far as what we really use, Green's theorem for (11.5), Kahler's condition holding in the sense of distributions is sufficient for the purpose.

Because of (11.4) the operator Δ , which is elliptic in $\mathcal{B}$, is singular at the boundary, and this singularity of Δ intimately reflects the strict pseudo-convexity of $\mathcal{B}$. Unfortunately at the present time, the general theory of solutions of such singular elliptic operators is little developed. We shall therefore have to make do with some very rudimentary facts concerning Δ : Green's theorem, and the existence of a rather explicit approximate solution of Δ given by the following lemma.

__Lemma 2.__ $|\Delta(\lambda(z))^n| \leq C|\lambda(z)|^{n+1}$, _for_ z _near the boundary._

Proof. Since $\dfrac{\partial^2\lambda^n}{\partial z_i\partial\bar{z}_j} = n\lambda^{n-2}\left[(n-1)\lambda_{z_i\bar{z}_j} + \lambda\lambda_{z_i\bar{z}_j}\right]$ we can write in matrix form

$$(11.6) \qquad \frac{\partial^2\lambda^n}{\partial z_i\partial\bar{z}_j} = n\lambda^{n-2}\left(\begin{array}{c|c} (n-1)a^2/4 + O(\lambda) & O(\lambda) \\ \hline O(\lambda) & -\lambda Q_2 + O(\lambda^2) \end{array}\right)$$

Taking the product of this matrix with the matrix (11.4) gives us

$$(11.7) \qquad \{g^{ij}\}\cdot\left\{\frac{\partial^2\lambda^n}{\partial z_i\partial\bar{z}_j}\right\} = n\lambda^n\left(\begin{array}{c|c} n-1 + O(\lambda) & O(\lambda) \\ \hline O(\lambda) & -I + O(\lambda) \end{array}\right)$$

Four times the trace of (11.7) is $\Delta(\lambda^n)$, which is therefore $O(\lambda^{n+1})$ proving the lemma.

__Remark.__ Observe that for general p, $p \neq n$ we could only have $\Delta(-\lambda)^p = O(|\lambda|^p)$. The function $(\lambda(z))^n$ plays part of the role of $\log|z|$ when $n = 1$ and $\mathcal{B}$ is the unit disc.

It will be convenient at this point to make explicit the gradient and the volume elements with respect to our metric (11.3). In general the gradient is the vector field given by
$\nabla f = 2 \sum_{i,j} g^{ij} \left\{ \frac{\partial f}{\partial \bar{z}_i} \frac{\partial}{\partial z_j} + \frac{\partial f}{\partial z_j} \frac{\partial}{\partial \bar{z}_i} \right\}$. If F is holomorphic (i.e. $\frac{\partial F}{\partial \bar{z}_i} = 0$), the square norm of ∇F is given by

$$|\nabla F|^2 = 2 \sum_{i,j} g^{ij} \overline{\frac{\partial F}{\partial z_i}} \frac{\partial F}{\partial z_j} .$$

Now for each z near $\partial\mathcal{D}$ choose the coordinate system induced by the splitting $\mathbb{C}^n = N_z \oplus C_z$, described above, with $N_z = \{(z_1, 0 \ldots 0)\}$, and $C_z = \{(0, z_2, z_3 \ldots z_n)\}$. We write also
$|\nabla_1 F|^2 = \left|\frac{\partial F}{\partial x_1}\right|^2 + \left|\frac{\partial F}{\partial y_1}\right|^2 = 2\left|\frac{\partial F}{\partial z_1}\right|^2$; and

$|\nabla_{2,n} F|^2 = \sum_{j=2}^{n} \left|\frac{\partial F}{\partial x_j}\right|^2 + \left|\frac{\partial F}{\partial y_j}\right|^2 = 2 \sum_{j=2}^{n} \left|\frac{\partial F}{\partial z_j}\right|^2$. Then in view of (11.4) we have

$$|\nabla F|^2 \approx \delta^2(z)|\nabla_1 F|^2 + \delta(z)|\nabla_{2,n} F|^2 . \tag{11.8}$$

The volume element $d\Omega$ given by the metric is

$$d\Omega = \det(g_{ij}) d\omega$$

where $d\omega = dx_1 \ldots dx_n dy_1 \ldots dy_n$ is the Euclidean element of volume. Since $\det(g_{ij}) \approx (-\lambda)^{-n-1} \approx \delta^{-n-1}$, we then have

$$d\Omega(z) \approx \delta(z)^{-n-1} d\omega(z) . \tag{11.9}$$

One final remark. It matters little which particular characterizing function λ, or resulting preferred metric we use. We keep one choice of λ fixed through §12 below.

12. The area integral and the local Fatou theorem

The main theorem of this chapter is as follows.

Theorem 12. *Let $\mathcal{D}$ be a domain in $\mathbb{C}^n$ with C^2 boundary which is strictly pseudo-convex. Suppose $F(z)$ is holomorphic in $\mathcal{D}$. Then for almost every $\zeta \in \partial\mathcal{D}$ the following three properties are equivalent:*

(a) F *is admissibly bounded at* ζ.

(b) F *has an admissible limit at* ζ.

(c) $\int_{\mathcal{A}_\alpha(\zeta)} |\nabla F|^2 \, d\Omega(z) < \infty$.

With the elements of the potential theory discussed in the previous section, the proof of this theorem follows the same lines as that for the known case of harmonic functions (in the usual potential theory) in a half-space in $\mathbb{R}^N$, as in Stein [19]. There are however certain interesting differences, and the main thrust of this section is to elucidate the modifications that are necessary.

We begin with three small lemmas. $\mathcal{A}_\alpha$ and $\mathcal{A}_\beta$ are two admissible approach regions (with common vertex, say $\zeta = 0$) and $\beta > \alpha$; thus $\mathcal{A}_\alpha \subset \mathcal{A}_\beta$. Notice also that if α is fixed there is a nontangential approach region Γ_γ (again with the same vertex) so that $\Gamma_\gamma \subset \mathcal{A}_\alpha$. F is assumed to be holomorphic in $\mathcal{A}_\beta$.

Lemma 1. Suppose $|F(z)| \leq C$ for $z \in \mathcal{A}_\beta$. Then

$$|\nabla F(z)| \leq AC \text{ for } z \in \mathcal{A}_\alpha.$$

Lemma 2. If $\int_{\mathcal{A}_\beta} |\nabla F|^2 d\Omega(z) < \infty$, then

$$|\nabla F(z)| \longrightarrow 0 \text{ for } z \in \mathcal{A}_\alpha, \quad z \longrightarrow 0.$$

Lemma 3. $\int_{\Gamma_\gamma} [|\nabla_1 F|^2 + |\nabla_{2,n} F|^2](\delta(z))^{2-2n} d\omega(z) \leq A \int_{\mathcal{A}_\beta} |\nabla F|^2 d\Omega(z)$.

Observe that $|\nabla F|$ is the length of the gradient of F taken with respect to the metric introduced in the previous section, and $|\nabla_1 F|^2 + |\nabla F|^2_{2,n}$ is the square of the Euclidean gradient (see also (11.8). Therefore the integral on the left side of the inequality of lemma 3 is the standard area integral discussed in §6.

To prove these lemmas we must begin by avoiding a pitfall. The Euclidean gradient has been split in two parts as $|\nabla_1 F|^2 + |\nabla_{2,n} F|^2$, but this splitting varies (with the decomposition $\mathbb{C}^n = N_z \oplus C_z$) as z varies in $\mathcal{A}_\beta$. We wish to make our calculations here with respect to a fixed choice of coordinates, and not this variable system. It is best then to take the coordinates which arise at the vertex $(z=0)$ of $\mathcal{A}_\beta$. Let us call these coordinates $(\tilde{z}_1, \tilde{z}_2, \ldots, \tilde{z}_n)$ and the corresponding parts of the gradients $|\tilde{\nabla}_1 F|^2 = 2\left|\frac{\partial F}{\partial \tilde{z}_1}\right|^2$,

$|\tilde{\nabla}_{2,n} F|^2 = 2 \sum_{k=2}^{n} \left|\frac{\partial F}{\partial \tilde{z}_k}\right|^2$. We need to observe that

(12.1) $\delta^2(z)|\nabla_1 F(z)|^2 + \delta(z)|\nabla_{2,n} F(z)|^2 \approx$

$$\delta^2(z)|\tilde{\nabla}_1 F(z)|^2 + \delta(z)|\tilde{\nabla}_{2,n} F(z)|^2$$

for $z \in \mathcal{A}_\beta$.

The only non-trivial assertions made by (10.10) are the statements that

$$|\tilde{\nabla}_{2,n}F(z)|^2 \leq C\{\delta(z)|\nabla_1 F(z)|^2 + |\nabla_{2,n}F(z)|^2\} ,$$

and the analogue with $\tilde{\nabla}$ and ∇ interchanged. Now $\frac{\partial F}{\partial \tilde{z}_k} = \sum_\ell a_k^\ell(z)\frac{\partial F}{\partial z_\ell}$, and if $z = 0$ then $a_k^\ell(0) = \delta_k^\ell$, since there $\frac{\partial}{\partial \tilde{z}_k} = \frac{\partial}{\partial z_k}$. Also the coefficients a_k^ℓ are clearly C^1 , so that $|a_k^\ell(z) - \delta_k^\ell| \leq C|z|$. But $|z|^2 < \beta(\delta(z))$ if $z \in \mathcal{A}_\beta(0)$, and inserting this above proves the statement (12.17).

As in §§9 and 10, we let $P(z)$ denote the polydisc centered at z , whose radii are essentially $c\delta(z),(c\delta(z))^{\frac{1}{2}},\ldots,(c\delta(z))^{\frac{1}{2}}$, with c sufficiently small. (Here the variables are split according to the decomposition $\mathbb{C}^n = N_\zeta \oplus C_\zeta$, where $\zeta = 0$.) Then $P(z) \subset \mathcal{A}_\beta$, if $z \in \mathcal{A}_\alpha$. Also

$$(12.2) \qquad |\tilde{\nabla}_1 F(z)| \leq c\delta^{-1} \sup_{z' \in P(z)} |F(z')|$$

$$(12.2)' \qquad |\tilde{\nabla}_{2,n}F(z)| \leq c\delta^{-\frac{1}{2}} \sup_{z' \in P(z)} |F(z')|$$

and so lemma 1 follows immediately upon applying (12.1). To prove lemma 2 we observe that

$$(12.3) \qquad |\tilde{\nabla}_1 F(z)|^2 \leq \frac{1}{|P(z)|} \int_{P(z)} |\tilde{\nabla}_1 F(z')|^2 \, d\omega(z')$$

$$(12.3)' \qquad |\tilde{\nabla}_{2,n}F(z)|^2 \leq \frac{1}{|P(z)|} \int_{P(z)} |\tilde{\nabla}_{2,n}F(z')|^2 \, d\omega(z')$$

and lemma 2 then follows also upon observing that $\delta(z') \approx \delta(z)$ if $z' \in P(z)$.

Now let $X(z,z')$ be the characteristic function of the set where $z' \in P(z)$. We observe that $|P(z)| \approx (\delta(z))^{n+1}$, and so by (12.3),

$$(12.4) \qquad \int_{\Gamma_\gamma} |\tilde{\nabla}_1 F(z)|^2 (\delta(z))^{2-2n} \, d\omega(z) \le$$
$$C \int_{\Gamma_\gamma} \int_{\mathcal{A}_\beta} (\delta(z))^{-3n+1} X(z,z') |\tilde{\nabla}_1 F(z')|^2 \, d\omega(z) d\omega(z') .$$

We have already remarked that $\delta(z) \approx \delta(z')$, if $z' \in P(z)$. We also note that

$$\int_{\Gamma_\gamma} X(z,z') d\omega(z) \approx (\delta(z'))^{2n} .$$

The left side of (12.4) is thus majorized by

$$C \int_{\mathcal{A}_\beta} (\delta(z'))^{-n+1} |\tilde{\nabla}_1 F(z')|^2 d\omega(z') \le C \int_{\mathcal{A}_\beta} |\tilde{\nabla}_1 F|^2 (\delta(z'))^2 d\Omega(z')$$
$$\le C \int_{\mathcal{A}_\beta} |\nabla F|^2 d\Omega(z') ,$$

because of (11.8), (11.9) and (12.1). One proves similarly that

$$\int_{\Gamma_\gamma} |\nabla_{2,n} F|^2 (\delta(z))^{2-2n} d\omega(z) \le A \int_{\mathcal{A}_\beta} |\nabla F(z)|^2 d\Omega(z) ,$$

which disposes of lemma 3.

We now prove that for almost every $\zeta \in \partial \mathcal{D}$, (a) ===> (c). In order to do this we can reduce matters to the following situation. E is a measurable subset of $\partial \mathcal{D}$, of sufficiently small diameter,

$F(z)$ is holomorphic and $|F(z)| \leq 1$, for $z \in \mathcal{A}_\beta(\zeta)$, whenever $\zeta \in E$, where β is fixed. It suffices to show that if $\alpha < \beta$ (α is also kept fixed), then for almost every $\zeta \in E$,

$$\int_{\mathcal{A}_\alpha(\zeta)} |\nabla F(z)|^2 d\Omega(z) < \infty ,$$

which is itself a consequence of

$$(12.5) \qquad \int_E \left\{ \int_{\mathcal{A}_\alpha(\zeta)} |\nabla F(z)|^2 d\Omega(z) \right\} d\sigma(\zeta) < \infty .$$

Let us prove (12.5). We write $\mathcal{R} = \bigcup_{\zeta \in E} \mathcal{A}_\alpha(\zeta)$, and $\psi(z,\zeta)$ for the characteristic function of the set $\{z \in \mathcal{A}_\alpha(\zeta)\}$. Then it is easy to see that $\int \psi(z,\zeta) d\sigma(\zeta) \leq c(\delta(z))^n$, and so (12.5) is a consequence of

$$(12.6) \qquad \int_{\mathcal{R}} (\delta(z))^n |\nabla F(z)|^2 d\Omega(z) < \infty .$$

Let us examine the region $\mathcal{R}$ more carefully. For simplicity of notation assume the origin belongs to E, and that the splitting $\mathbb{C}^n = N_0 \oplus C_0$ corresponding to this boundary point is the standard one, where $N_0 = \{(z_1, 0, \ldots, 0)\}$, $C_0 = \{(0, z_2, z_3, \ldots, z_n)\}$, and $\nu_0 = (-i, 0, \ldots, 0)$ is the unit outward normal at 0. The other points $\zeta \in E$ have the property that $|\nu_0 - \nu_\zeta| < c$, where c can be chosen to be small. Now suppose Γ is a small truncated cone whose vertex is at the origin and whose axis is directed along γ_0: $\Gamma = \{z: |z_2|^2 + \ldots |z_n|^2 + x_1^2 < \bar{c} y_1^2$ and $0 < y_1 < \bar{c}\}$, where $\bar{c}$ is small. Write $\Gamma_+(z) = \{\Gamma + z\}$ and $\Gamma_-(z) = \{-\Gamma + z\}$. We also divide the boundary of $\mathcal{B} = \partial\mathcal{R}$ into two parts $\mathcal{B} = \mathcal{B}_0 \cup \mathcal{B}_1$. For fixed c_1 (which is chosen small), $\mathcal{B}_0 = \{z \in \mathcal{B}, \delta(z) < c_1\}$ and $\mathcal{B}_1 = \{z \in \mathcal{B}, \delta(z) > c_1\}$. Thus $\mathcal{B}_1$ is the "trivial" part of the boundary, lying strictly in the interior of $\mathcal{D}$. $\mathcal{B}_0$ is the critical part of the boundary of $\mathcal{R}$. Our claim is that $\mathcal{B}_0$ is a part of a Lipschitz hypersurface given by $y_1 = \Psi(x_1, z_2, \ldots, z_n)$, with

$$|\Psi(x_1, z_2, \ldots z_n) - \Psi(x_1', z_2', \ldots z_n')| \leq M\left\{|x_1 - x_1'| + \sum_{k=2}^{n} |z_k - z_k'|\right\}.$$

To see this, it suffices to observe that each point $z \in \mathcal{B}_0$ has the following property: the forward-cone $\Gamma_+(z)$ lies in $\mathcal{R}$, and the backward cone $\Gamma_-(z)$ lies in ${}^c\mathcal{R}$. In fact each point z of $\mathcal{B}_0$ is on the boundary of some $\mathcal{A}_\alpha(\zeta)$, for $\zeta \in E$. It is also clear that for such z, $\Gamma_+(z) \subset \mathcal{A}_\alpha(\zeta)$, and $\Gamma_-(z) \subset {}^c\mathcal{A}_\alpha(\zeta)$ if $z \notin \mathcal{A}_\alpha(\zeta)$, the constants c, $\bar{c}$, and c_1 are sufficiently small.

Thus (as in [19], chapter VII), we may approximate $\mathcal{R}$ by a family $\mathcal{R}_\varepsilon$ of smooth regions so that their boundaries can be decomposed as $\partial\mathcal{R}_\varepsilon = \mathcal{B}_\varepsilon^0 + \mathcal{B}_\varepsilon^1$. Here if $z \in \mathcal{B}_\varepsilon^1$, then $\delta(z) > c_1/2$ while $\mathcal{B}_\varepsilon^0$ is part of a smooth hypersurface, at positive distance from $\partial\mathcal{R}$, with a Lipschitz constant that may be taken independent of ε. Also $\bar{\mathcal{R}}_\varepsilon \subset \mathcal{R}$ and $\mathcal{R}_\varepsilon \nearrow \mathcal{R}$.

To prove (12.6) it suffices to show therefore that

$$(12.7) \qquad \int_{\mathcal{R}_\varepsilon} (\delta(z))^n |\nabla F(z)|^2 d\Omega(z) \leq c < \infty .$$

We now use Green's formula (for our metric (11.3)) in terms of the Laplace-Beltrami operator (11.5). This formula is

$$(12.8)\qquad \int_{\mathcal{R}_\varepsilon}\left[A\Delta B - B\Delta A\right]d\Omega = \int_{\mathcal{B}_\varepsilon}\left[A\frac{\partial B}{\partial n_\varepsilon} - B\frac{\partial A}{\partial n_\varepsilon}\right]d\tau_\varepsilon .$$

Here $\frac{\partial}{\partial n_\varepsilon}$ are normal derivatives in the direction of the outward unit normal, but with length normalized according to the metric (11.1). $d\tau_\varepsilon$ is the measure on $\mathcal{B}_\varepsilon$ induced by the element of volume given by the metric; that is $d\tau_\varepsilon$ is induced by $d\Omega$.

To apply the formula (12.8) we use the fact that $\Delta|F|^2 = 2|\nabla F|^2$, when F is holomorphic, since the metric is Kahler (see formula (11.5), and the one following (11.7)). Also $(\delta(z))^n \approx (-\lambda(z))^n$, while $|\Delta\lambda^n| \leq C|\lambda^{n+1}|$, by lemma 2 of §11. Substituting $A = (-\lambda)^n$, $B = |F|^2$ in (12.8) we therefore get that the left side of (12.7) is majorized by a constant multiple of

$$(12.9)\qquad \int_{\mathcal{B}_\varepsilon}(-\lambda)^n\frac{\partial|F|^2}{\partial n_\varepsilon}d\tau_\varepsilon - \int_{\mathcal{B}_\varepsilon}|F|^2\frac{\partial(-\lambda)^n}{\partial n_\varepsilon}d\tau_\varepsilon + \int_{\mathcal{R}_\varepsilon}|\lambda|^{n+1}|F|^2 d\Omega .$$

Now since $d\Omega \approx \delta^{-n-1}d\omega \approx \lambda^{-n-1}d\omega$, we see that the last term is dominated by a constant multiple of

$$\int_{\mathcal{R}_\varepsilon}|F|^2 d\omega \leq \int_{\mathcal{R}}|F|^2 d\omega < \infty ,$$

because F is bounded in $\mathcal{R}$. To take care of the first two terms of (12.9) it suffices clearly to consider only the part of the boundary near $\partial\mathcal{R}$, namely $\mathcal{B}_\varepsilon^o$. We prove first

$$(12.10)\qquad -c_1|\lambda|^n \leq \frac{\partial(-\lambda)^n}{\partial n_\varepsilon} \leq -c_2|\lambda|^n,\quad c_1 > c_2 > 0,\quad \text{on } \mathcal{B}_\varepsilon^o$$

and

(12.11) $$\int_{\mathcal{B}^0_\varepsilon} |\lambda(z)|^n d\tau_\varepsilon \leq c < \infty .$$

Let ν_ε denote the unit outward normals (in the Euclidean lenth) at $\mathcal{B}^0_\varepsilon$. Because of the uniform Lipschitz character of the hypersurfaces $\mathcal{B}^0_\varepsilon$, the normals ν_ε have a positive projection along all the unit normals ν_ζ, for $\zeta \in \partial\mathcal{D}$, whenever ζ is sufficiently close to E, and E is sufficiently small. Since λ is a characterizing function of $\partial\mathcal{D}$, it then follows that

$$\bar{c}_1 \geq \frac{\partial\lambda}{\partial\nu_\varepsilon} \geq \bar{c}_2 > 0 .$$

Again because $\nu_\varepsilon(z)$ has a positive projection with ν_ζ, where $\zeta = n(z)$ is the normal projection of z on $\partial\mathcal{D}$, it follows by the defining properties of the preferred metric that

$\dfrac{\partial(-\lambda)^n}{\partial n_\varepsilon} = \dfrac{\partial(-\lambda)^n}{\partial\nu_\varepsilon}$ times a positive factor of order of magnitude δ, which gives (12.10).

To prove (12.11) use Green's formula (12.8) again, this time with $A \equiv 1$, and $B = (-\lambda)^n$. Then since $|\Delta(-\lambda)^n| \leq A|\lambda|^{n+1}$, and (12.10) we get that

$$\int_{\mathcal{B}^0_\varepsilon} |\lambda|^n d\tau_\varepsilon \leq \int_{\mathcal{B}^1_\varepsilon} |\lambda|^n d\tau_\varepsilon + \text{constant x} \int_{\mathcal{R}_\varepsilon} |\lambda|^{n+1} d\Omega$$

and both terms on the right are trivially bounded. We return to (12.9). The boundedness of the third term has already been described. Since $|F| \leq 1$ in $\bigcup_{\zeta \in E} \mathcal{a}_\beta(\zeta) \subset \mathcal{R}$, and $\mathcal{B}_\varepsilon \subset \mathcal{R}$ the boundedness of the second term follows from (12.10) and (12.11). The boundedness of the first term proved similarly, once we apply lemma 1 of this section, which shows that $|\nabla F| \leq A$, and hence $\left|\frac{\partial |F|^2}{\partial n_\varepsilon}\right| \leq 2|F|\nabla F| \leq 2A$, since $|F| \leq 1$ in $\mathcal{A}_\beta(\zeta)$, $\zeta \in E$. This proves (12.7) and thus (12.6), concluding the proof that almost everywhere property (a) implies property (c).

We now prove that almost everywhere (c) ===> (b). (That (b) ===> (a) is trivial.)

Because of lemma 3 of this section, and theorem 5 of §6, we see that at almost all points where $\int_{\mathcal{A}_\beta(\zeta)} |\nabla F|^2 d\Omega(z) < \infty$, F has non-tangential limits. Now let ζ_o be any such point, and assume that it is the origin and the coordinates have been so chosen that $\{(z_1, 0 \ldots 0)\} = N_o$, and $\{(0, z_2, \ldots z_n)\} = C_o$. Write $F(z) = \{F(z) - F(z_1, 0, \ldots 0)\} + F(z_1, 0, \ldots 0)$. The first term on the right side is dominated by

$$|z(1) - z(0)| \sup_{0 \leq t \leq 1} |\tilde{\nabla}_{2,n} F(z(t))|, \text{ where } z(t) = (z_1, z_2 t, \ldots, z_n t).$$

If $z \in \mathcal{A}_\alpha(0)$, then $z(t) \in \mathcal{A}_\alpha(0)$, and there $\delta(z(t)) \approx \delta(z)$ while $|z(1) - z(0)| \leq c\delta(z)$. Thus by lemma 2 (together with (11.8) and (12.1)) we have that $F(z) - F(z_1, 0, \ldots 0) \longrightarrow 0$ as $z \longrightarrow 0$ in $\mathcal{A}_\alpha(0)$. Since $(z_1, 0, \ldots, 0) \longrightarrow (0, 0, \ldots, 0)$ non-tangentially (as $z \longrightarrow 0$ in $\mathcal{A}_\alpha(0)$), we conclude that $\lim_{\substack{z \in \mathcal{A}_\alpha(\zeta) \\ z \to \zeta}} F(z)$ exists at almost all ζ for which $\int_{\mathcal{A}_\beta(\zeta)} |\nabla F|^2 d\Omega < \infty$. Finally, at every point which is a point of density (with respect to the balls $B_2(\cdot, \rho)$)

of this set, the limit also exists when we allow arbitrarily large α as apertures for the approach regions $\mathcal{A}_\alpha(\zeta)$. The latter fact can be proved by a simple adaptation of the argument in the classical case (see e.g. Stein [19], p.202).

This completes the proof of theorem 12.

Remark. It is clear that argument (c) ===> (b) works without any assumption of strict pseudo-convexity, since the properties of the metric (11.1) are in reality not used there. The question arises: Is theorem 12, in its entirety, valid without any assumptions of pseudo-convexity?

References for Chapter III

The basic properties of the preferred metric, and theorem 12 were outlined in Stein [20]. For the case of the unit ball in $\mathbb{C}^n$, for which more can be said, see the forthcoming paper of Robert B. Putz, "The generalized area theorem for harmonic functions on Hermitian hyperbolic space."

Bibliography

[1] Aronszajn, N. and Smith, K.T., Functional spaces and functional completion, Ann. Inst. Fourier 6(1955), 725.

[2] Bergman, S., "The kernel function and conformal mapping", 2nd edition (1970), A.M.S. Survey n$^{\underline{o}}$ 5.

[3] -----, "Sur la function-noyau d'un domaine ...", Mém. Sci. Math. Paris, 108(1948).

[4] Gunning, R. C. and Rossi, H., "Analytic Functions of Several Complex Variables", Prentice Hall, 1965.

[5] Helgason, S., "Differential Geometry and Symmetric Spaces", New York, 1962.

[6] Hörmander, L., L^p estimates for (pluri-) subharmonic functions, Math. Scand. 20(1967), 65-78.

[7] -----, "An Introduction to Complex Analysis in Several Variables", Van Nostrand, 1966.

[8] Hua, L. K., "Harmonic Analysis of Functions of Several Complex Variables in the Classical Domains", A.M.S. (1963).

[9] Kohn, J. J., Harmonic integrals on strongly pseudo-convex manifolds I, Annals of Math. 78(1963), 112-148.

[10] Koranyi, A., Harmonic functions on Hermitian hyperbolic space, Trans. Amer. Math. Soc. 135(1969), 507-516.

[11] -----, The Poisson integral for generalized half-planes and bounded symmetric domains, Annals of Math. 82(1965), 332-350.

[12] Malliavin, P., Comportement à la frontière distinguée d'une fonction analytique de plusieurs variables, C.R.A. Sci. Paris, 268(1969), 380-381.

[13] Malliavin, P., Théorème de Fatou en plusieurs variables complexes, (to appear), preprint pp.47-53.

[14] Pjateskii, I. I.- Shapiro, "Geometry of Classical Domains and Automorphic Functions", Fizmatgiz (1961) (Russian).

[15] Privalov, I. I. and Kuznetzov, P. I., Boundary problems and various classes of harmonic and subharmonic functions defined in arbitrary regions, Nat. Sbornik 48(1939), 345-375 (Russian).

[16] Schiffer, M. and Spencer, D. C., "Functionals of finite Riemann surfaces, Princeton, 1954.

[17] Siegel, C. L., "Analytic Functions of Several Complex Variables", Inst. for Advanced Study, 1950.

[18] Smith, K. T., A generalization of an inequality of Hardy and Littlewood, Canad. J. Math. 8(1956), 157-170.

[19] Stein, E. M., "Singular Integrals and Differentiability Properties of Functions", Princeton (1970).

[20] -----, Boundary values of holomorphic functions, Bull. Amer. Math. Soc. 76(1970), 1292-1296.

[21] -----, The analogues of Fatou's theorem and estimates for maximal functions, in "Geometry of Homogeneous Bounded Domains", C.I.M.E., 1967.

[22] ----- and Weiss, G., "Introduction to Fourier Analysis on Euclidean Spaces", Princeton, 1971.

[23] ----- and Weiss, N. J., On the convergence of Poisson integrals, Trans. Amer. Math. Soc. 140(1969), 35-54.

[24] Weil, A., Introduction à l'étude des rariétés Kählériennes, Hermann, Paris, 1958.

[25] Widman, K.-O., On the boundary behavior of solutions to a class of elliptic partial differential equations, Ark. Mat. 6(1966), 485-533.

[26] Zygmund, A., Trigonometric Series, Cambridge, 1959.

Lightning Source UK Ltd.
Milton Keynes UK
UKHW022041280820
368995UK00006B/79

9 780691 620114